The Broken Machine

The Broken Machine

Histories of Technology, Social Order, and the Self

Edward Jones-Imhotep

The MIT Press
Cambridge, Massachusetts
London, England

The MIT Press
Massachusetts Institute of Technology
77 Massachusetts Avenue
Cambridge, MA 02139
mitpress.mit.edu

The MIT Press would like to thank the anonymous peer reviewers who provided comments on drafts of this book. The generous work of academic experts is essential for establishing the authority and quality of our publications. We acknowledge with gratitude the contributions of these otherwise uncredited readers.

This book was set in Stone Serif and Stone Sans by Westchester Publishing Services. Printed and bound in the United States of America.

Library of Congress Cataloging-in-Publication Data

Names: Jones-Imhotep, Edward author
Title: The broken machine : histories of technology, social order, and the self / Edward Jones-Imhotep.
Description: Cambridge, Massachusetts : The MIT Press, [2026] | Includes bibliographical references and index.
Identifiers: LCCN 2025032227 (print) | LCCN 2025032228 (ebook) | ISBN 9780262553346 paperback | ISBN 9780262384704 pdf | ISBN 9780262384711 epub
Subjects: LCSH: Inventions—History | New products—History | System failures (Engineering)—History | Business failures—History
Classification: LCC T15 .J66 2026 (print) | LCC T15 (ebook)
LC record available at https://lccn.loc.gov/2025032227
LC ebook record available at https://lccn.loc.gov/2025032228

EU Authorised Representative: Easy Access System Europe, Mustamäe tee 50, 10621 Tallinn, Estonia | Email: gpsr.requests@easproject.com

For Inés, Xavier, and Tamara

Contents

Contents

Acknowledgments

I have carried this book with me for many years, and through events I couldn't have imagined. That it is appearing in the world is a testament to the many people, some of them complete strangers, who have helped make it real. It is a pleasure to thank them here.

Researching and writing this book would have been impossible without generous financial and institutional support from many sources. Grants from York University, the Social Sciences and Humanities Research Council of Canada, and the Deutscher Akademischer Austausch Dienst (DAAD) supported the archival research at the book's core. A Jackman Humanities Institute Fellowship and an administrative leave from the University of Toronto in 2023-2024 allowed me valuable time away from my role as Director of the Institute for the History and Philosophy of Science and Technology (IHPST) for writing and reflection. My deep thanks to the incomparable Denis Walsh for taking over as Acting Director that year. And although it may be rare to thank university administrators in acknowledgements, I owe an enormous debt to Dean Melanie Woodin (now President of the University of Toronto) for her unstinting support at every stage.

When doing anything that takes time and imagination, Hanif Abdurraqib reminds us, it behooves you to have a crew. I have been blessed to have so many people who offered their time, friendship, support, and rich insights as this book unfolded: Jennifer Bonnell, Stephen Brooke, Libby Cohen, Tom Cohen, Tina Choi, Boyd Cothran, Jonathan Edmonson, Paul Edwards, Josh Fogel, Sakis Gekas, Gabrielle Hecht, Steve Jackson, Eric Jennings, Joan Judge, Michelle Johnson, Sean Kheraj, David Koffman, Rachel Koopmans, Molly-Ladd Taylor, Paul Lawrie, Marcel Martel, Massimo Mazzotti, Mary Mitchell, M. Murphy, Deborah Neill, Sharrona Pearl,

Sara Pritchard, Jim Retellack, Scott Richmond, Ann Rubenstein, Margaret Schotte, Suman Seth, Asif Siddiqi, Rebecca Slayton, Alison Smith, Jennifer Stephen, Emily Thompson, Heidi Voskuhl, and Bill Wicken. I am forever grateful to Cheryl Misak for bringing me to the University of Toronto, and for all her warmth, mentorship, and encouragement along the way. I owe her more than I can express.

In the depths of the pandemic, I was fortunate enough to find a remarkable writing group. Week after week, their wisdom, humor, generosity, and brilliance helped me carry not only the weight of this book, but so much else. I am constantly reminded of how lucky I am that they came into my life when they did. My deep thanks to the group's wise and fearless leader, Therí Pickens, to Crystal Lucky and Lora Levitt, and most of all to Julia Elyachar, who has continued to be a source of strength, compassion, friendship, and inspiration for me ever since.

I am grateful as well to audiences at the ETH Zurich, Centre Alexandre Koyré, École des Hautes Études en Science Sociales (EHESS), History of Science Society, Society for the History of Technology, University of Oklahoma, University of British Columbia, University of California—Berkeley, University of Leeds, University of Michigan, University of Minnesota, University of Toronto, and Yale University for their feedback. A special thanks to Charlotte Bigg, Bob Brain, John Carson, Debbie Coen, Henry Cowles, Lucia Dacome, Anna Garber, Graeme Gooday, Leslie Higgins, Michel Janssen, Paul Johnson, Nikolai Krementsov, Carla Nappi, David Nye, Stephan Palmié, Greg Radick, Bill Rankine, Iwan Rhys Morus, Joy Rohde, Catherine Rosenthal, Suzanne Moon, Pete Soppelsa, Dagmar Schäfer, Simon Schaffer, and members of the Victorian Studies Network at York University for their generous comments on earlier drafts of the material.

At every stage of working on this book, people selflessly took the time to read portions and offer encouragement, suggestions, and helpful criticism. They include Tina Choi, Joy Rohde, Hunter Heyck, Jamie Cohen-Cole, Ken Lipartito, Alicia Maggard, Taylor Moore, Nicolas Nova, Mike Pettit, and students and faculty at the Zentrum Geschichte des Wissens, ETH Zürich, especially Silvia Berger, Kijan Espahangizi, Manuel Kaiser, Patrick Killian, Daniel Strassberg, and Vera Wolff. A special thanks to four people who read the entire manuscript. Matthew Farish, Mike Pettit, Tiago Saraiva, and Rebecca Woods have been cherished colleagues for as long as I can remember. Their generosity and brilliance make it a joy to think with them. Finally, thanks to the anonymous referees who provided invaluable feedback and comments at various stages. I'm deeply grateful for their time, encouragement, generosity, and enthusiastic support.

The staff of many archives, libraries, and museums opened up the historical materials that are at the heart of this book. I am especially grateful to Doug McQuirter at the Hagley Museum and Library; Bethany Aitchison and David Pantalony at the Canada Science and Technology Museum; Linda Beninghove and Ted Houghtaling of the Archives and Special Collections at Stevens Institute of Technology; Lise Cloitre, Reisse Chrystèle, and Florence Desnoyers-Robison at the Conservatoire National des Arts et Métiers, Paris; Skip Vaughn at the US Army Garrison Public Affairs Office; and archivists at Library and Archives Canada and at the British National Archives, Kew.

With help preparing the manuscript, I'm indebted to research assistants Myles Ali, Ari Finsson, Nayani Jensen, and Birch Norman. A special thanks to the incomparable Bree Lohman, who helped track down permissions and prepare the final submission. Natalie Tomlin and Leslie Castro Woodhouse provided valuable editorial advice on the entire manuscript. I would also like to thank staff at the MIT Press, including Katie Helke, Suraiya Jetha, and Roger Wood. To my editor, Justin Kehoe, I cannot repay the great gift of your patience, time, and guidance throughout this journey.

Earlier and substantially different versions of some chapters of this work appeared in previous publications. Material from Chapter 1 appeared in "The Unfailing Machine: Mechanical Arts, Sentimental Publics, and the Guillotine in Revolutionary France," *History of the Human Sciences* 30, No. 4 (2017): 11-31. Much earlier versions of material in Chapter 5 appeared in "Icons and Electronics." *Historical Studies in the Natural Sciences* 38 (2008): 405-450; and "Maintaining Humans: Electronic Failure and Human Nature," in *Cold War Social Science: Knowledge Production, Liberal Democracy, and Human Nature*, edited by Mark Solovey and Hamilton Cravens, 225-243 (New York: Palgrave Macmillan, 2012). I am very grateful for permission to reproduce them here.

Finally, and always, loving thanks to my entire family, immediate and extended, near and far, past and present. I would not be who I am without all of you. My wife and children already know that my deepest debts of all are always to them: to Tamara, for her inexhaustible encouragement, support, patience, and good humor; and to Xavier and Inés, who inspire me every day without knowing it, and who make everything worthwhile. Nothing in my life can compare to the joy you all bring. I dedicate this book to you.

Introduction

Breakdowns are portals into worlds we thought we knew.

It was never supposed to break. Of all the machines at work in Revolutionary Paris in the summer of 1792—clocks and optical telegraphs, presses and winches, weaving looms and pile drivers—this one device was supposed to be infallible. The surgeon Antoine Louis, France's leading expert on death, had been very clear on this point when he designed the ten-foot mechanism out of tempered steel, solid oak, and thick rope: *It must not fail.* The revolution depended on it. Murders and crimes were multiplying. National honor was at stake. The Enlightenment's highest ideals of justice and humanity demanded nothing less. But this was the last execution of the day and the machine had been built hastily. The weather had been warm and stormy during the week of July 16 and the thick wooden beams had drunk in the moisture. And so, in full view of the throng of spectators gathered to see the two-month old machine in the bustling Place de Grève, in the heart of Paris, the guillotine's blade jammed inside its wooden guideposts partway through its terrible action.

A few blocks away from the macabre scene, in the posh government offices of the Place Vendôme, news of the breakdown soon reached the *procureur-général-syndic,* Pierre Louis Roederer. He had commissioned the machine's construction. As the man in charge of local law enforcement and public order, Roederer was passing through dark days. Just weeks earlier, a crowd of ten thousand *sans coulottes* had marched on the Legislative Assembly and then on the Tuileries Palace to confront the King, forcing him to wear the red cap of liberty and drink a toast to the people and the nation. Public order, it seemed, rested on a knife-edge. Now this.

What happened next was remarkable. But before we return to the public theater of breakdown and its unfolding drama, ask yourself this: how should we think about this moment of breakage in one of history's most iconic technologies? Here are two possibilities. First, you might see the guillotine's breakdown as an aberration—an unfortunate but rare technological glitch in the workings of a machine that, if anything, worked *too* well. On this view, the world is filled with normally functioning technologies whose workings underpin and shape the societies they inhabit. The guillotine was no exception. Its proper operation embodied France's break with the cruelty of the Old Order and the nation's newly enlightened commitments to equality, humanity, and justice. A breakdown threatened all this. But like most breakdowns, you may be thinking, that episode was fleeting and of little consequence. What happened in the Place de Grève that July afternoon ultimately did little to stop the machine's brutal efficacy, as it transformed the history of punishment, enforced the successive social visions that spiralled through the French Revolution, and turned the guillotine itself into a powerful fixture of the modern medical, scientific, judicial, social, and cultural imagination. Here the working machine is what matters. Because only it teaches us how, in Bruno Latour's words, "technology is society made durable."[1] We can think of this as the *working-world* view of technology, and it has dominated not only histories of the guillotine but histories of technology more generally.[2]

Our other option is to see the breakdown itself as the default—one of the inescapable entropies that constantly swirl around technologies, tormenting them like furies. Here, you might say, the guillotine simply succumbed to the decadent tendencies in all machines, the ones held at bay by careful maintenance or reversed by quick repair. If the *working-world* view sees machines and social orders as stable and self-sustaining, this view casts machines and social relations as fragile.[3] As moments of unpredictability and script-breaking, breakdowns are everywhere. They represent dangerous ruptures in the social fabric, exposing the unspoken taboos, complex infrastructures, social inequalities, and legions of workers that sustain industrial societies but rarely figure in their histories.[4] Because breakdowns are everywhere, modern societies dedicate immense resources to fixing and preventing them. In doing so, they create the illusion of a "working world" by devaluing, ignoring, or hiding from view the acts of maintenance that sustain both their technologies and their social orders.[5] The mostly anonymous people who built, repaired, and maintained the guillotine stood in a long line of history's mostly invisible technicians, ignored when things worked and blamed when they broke.[6] On this view, rather than anomalies or banalities, breakdowns are brief but shining moments of revelation, making visible the subtle social relations, buried infrastructures, forgotten figures, hidden actions, and obscured dependencies that go unnoticed when technologies function as expected.[7] Borrowing Steve Jackson's beautiful phrase, we might call this the *broken-world* view of technology and it has inspired us to see our technological worlds in a new and important light.[8]

These two views—*working-world* and *broken-world*—might seem like opposites, but they converge on at least three crucial points. First, both believe that working technologies make for working societies. Paraphrasing Simon Schaffer and Steven Shapin's famous dictum about knowledge, they rest on a belief that solutions to the problem of technology (including breakage) are solutions to the problem of social order.[9] Second, because of this, both views cast breakdowns themselves as suspensions or interruptions to the social order, either trivial and inconvenient (in the *working-world* view) or significant but constantly mended (in the *broken-world* view). Finally, in both cases—either because they are largely irrelevant or because they are quickly repaired—breakdowns themselves, as objects, quickly disappear from the story. Working machines, or machines made to work again, are what matter. It is as if breakdowns make us uncomfortable—so uncomfortable that we cannot sit with them for very long.

But what if, rather than giving into that impulse in ourselves, we turned it into an object of historical investigation? What if we saw our own discomfort with breakdowns as part of a learned response, culturally reinforced over centuries, and serving specific political motivations? What would happen to our histories of technology if we resisted the temptation to move past the broken machine—resisted it for the length of time, say, that it took to read a book—and instead sat with what Nella Larsen called the "disturbing element"?[10]

The Broken Machine takes up these questions. It proposes a third possibility for thinking about the relationship between modern machines, breakdowns, and social order.

Consider once again the events surrounding the guillotine in the Place de Grève in July 1792. What I have always found remarkable about this little-reported episode is what historical actors did *not* say in this terrible moment. An execution had just gone horribly wrong. The old political order lay in ruins. France was at war. Crimes were multiplying. Foreign enemies threatened invasion. Traitors and royalists plotted from within. If either the *working-world* or the *broken-world* views gave a full account of the kinds of events breakdowns represent, then this would be the moment for officials to stress the importance of a working guillotine for the revolutionary social order and the urgency of getting the machine working again. Was the credibility of the enlightened nation damaged? Did the revolutionary crowds, never far from violence, demand justice or revenge? Was the fragile social order of the late eighteenth century plunged once again into peril?

But French officials and their powerful allies asked none of these questions, at least not outwardly. Instead, their anxiety focused on the question of the breakdown itself. Who was to blame? Who had the authority to talk about the event, its causes, its meanings? And perhaps most importantly, who was troubled—who was *disturbed*—by a breaking machine? On that summer afternoon in 1792, a mechanical breakdown became neither a curious sideshow nor a momentary but fleeting material and social rupture. It became an expanded microcosm for articulating visions of social order, identity, and belonging.

For a week afterward, it consumed the passions and paper and ink of some of the most powerful figures in revolutionary Paris—magistrates, royal architects, the king's carpenter, beleaguered ministers. It made officials' blood boil. It competed with the Haitian Revolution for space in letters from the finance minister, Clavière, or with the besieged monarchy in Roederer's urgent appeals. Most importantly for this book, it became an instrument for evaluating the qualities of the populace and, more particularly, the selves—the *kinds of people*—inhabiting revolutionary France: sentimental spectators, untrustworthy foreigners, loyal functionaries, authoritative architects. All were defined in some way through their attitudes, real and imagined, to the breakdown. In those turbulent moments, Roederer and his powerful colleagues enrolled not just the working guillotine but the broken machine in the project of revolutionary governance and their sweeping, anxious visions of French society.

The story of the guillotine is part of a larger and fascinating history of the politics of breakdown in the modern period. *The Broken Machine* explores selected elements of that larger history. You may feel that you're already familiar with its broad contours. There is a well-known politics surrounding technological disaster, for instance, where modern political elites have routinely turned the calamities of machines into opportunities to enact sweeping social visions, imposing questionable policies on shocked populations.[11] This large-scale, top-down reshaping of society after technological catastrophes is not what I have in mind, though. Similarly, there is a well-studied politics of maintenance and repair, where breakdowns, for example, formed moments of potential danger in the scientific, industrial, and political revolutions of the late eighteenth and nineteenth centuries, precisely because mending machines and mending social relations were of a piece.[12] But this small-scale, after-the-fact restoration of techno-social orders is not what I have in mind either. As valuable as those insights are, they don't help us understand the longer, varied, and more complicated histories of breakdowns as cultural and political objects.

From the late eighteenth century onward, in an increasing number of locations around the globe, what you felt in the face of a breakdown—panic, disquiet, rapture, indifference—said something about your *self*. Using breaking machines that ranged from steam engines to spinning gins, steel presses to atomic weapons, a collection of modern social elites—politicians, physicians, engineers, social reformers, military strategists—believed that one's experience of breakdowns gave insights into their soul. It is as if they proffered an answer to a powerful riddle: "Tell me how you feel about a broken machine, and I will tell you who you are." They interrogated their subjects through increasingly sophisticated instruments: tables, punishment books, management charts, and automated machines. They weighed their results against an increasingly rigid belief: that proper people, people belonging to the modern social order, should be troubled by the failures of machines. Their visions of breakage and social order, and their increasingly ambitious attempts to make people feel the failures of machines and to make breakdowns a problem of the modern self, are the chief concern of this book. Through their efforts, being disturbed by breakdowns came to seem like the proper and self-evident response to these events. If it seems that way to you, it is likely because you are part of the histories told in these pages.

Our theories of technology have done so much to explain what it means for technologies to *work*, or to *work again*, that we are unused to thinking about the work done by broken machines.[13] But beyond the working machine lies another history of technology—one buried beneath an attitude to breakdowns that makes them seem collectively too fleeting, mundane, and inconsequential to deserve sustained historical attention. Although working machines may be put to myriad uses, broken machines are seemingly useless, especially to historians exploring the powerful and formative currents of the past.[14] In writing about the Haitian Revolution, Michel-Rolph Trouillot, however, called our attention to the "formulas of erasure" and "formulas of banalization" that shape historical writing—the scholarly practices that erase or trivialize consequential historical developments to create the silences and forgotten ruins of our historical past.[15] The history of technology has seen its share of these formulas. Breakdowns have been one of their casualties. But as Trouillot shows, discarded objects sometimes hold the most important lessons of all. In our own time, we are in desperate need of new ways to understand the machines that surround us, their shortcomings, malfunctions, and failures, their relation to our social worlds, and what their histories, in all their complexity, say about us.

This book takes up that challenge through five historical narratives of breakdowns, exploring how breaking and broken machines figured in political projects and social visions from eighteenth-century political revolution to late twentieth-century global-scale warfare. Breakdowns represent the accidental elements of the history of technology—not in the sense of disasters or misfortunes, although they sometimes take those forms, but rather as suspensions of the regular and habitual order of things. Perhaps because of this, within historical writing they occupy the role accidents themselves did until recently: not so much overlooked but rather marginalized as historically significant sites of power and contested meanings and as rich, evolving, collective historical and social phenomena in their own right.[16] I focus on how government officials, medical professionals, elites, and social reformers used the breakdowns of their age to articulate and define the kinds of people who were central to their functioning societies: the sentimental self of the French Revolution, the responsible self of Victorian Britain, the efficient self of Progressive Era America, and the reasoning and rational self of the North American Cold War. Temporally, the book covers the period from the French and Haitian Revolutions in the late eighteenth century to the decolonization movements and the Cold War in the twentieth century. Geographically, it focuses on the North Atlantic: France, Britain, and their former North American colonies. Within that bounded time and place, it shows how histories of breaking machines might illuminate a politics of breakdown that has helped shape both our technologies and ourselves.

Breakdowns and the Objects Without History

Almost a century ago, the architectural critic, urban theorist, literary scholar, and historian of technology, Lewis Mumford, declared that breakdowns mattered little to history. He made that pronouncement in his epic *Technics and Civilization*—one of the foundational texts in the history of technology. Mumford was trying to unravel one of the most tortured questions in technology studies: How do technologies figure in questions of social order? He thought he found an answer in the steady rhythms of modern machines, the first of which (he believed) was the medieval clock. The unfailing cadence of the clock gave human society "the regular collective beat and rhythm of the machine; for the clock is not merely a means of keeping track of the hours, but of synchronizing the actions of men."[17] Over the thousand-year history that followed—across continents and ages, through watermills, steam engines, and combustion chambers—Mumford dedicated only a single page to breakdowns. On that single page, he drove a deep wedge between them and the making of modern peoples, arguing that modernity itself was built on their estrangement. Breakdowns held a single lamentable significance for Mumford: they had become irrelevant. Modernity's "machine-trained populations" were defined by their *indifference* to them.[18]

The Broken Machine shows how wrong Mumford was, how strange this view of technology is, and how much it hides. It argues that the very indifference Mumford lamented, the one he felt characterized modern peoples, was, in fact, a consistent and prominent *target* of European and North American elites from the eighteenth century onward. Over two centuries, a range of figures—medical officials, engineers, politicians, social reformers—cast breakdowns as a problem of the self: a problem of the proper relationship between breaking machines and inner lives. For these historical figures, being disturbed by breakdowns was a virtue: a moral quality that signaled a person's deep investment in functioning machines and therefore in the social order. Where they saw that quality lacking, they developed techniques to cultivate, nurture, and enforce it in some and to pathologize its absence in others. Their efforts turned breakdowns into what Barbara Cruikshank calls a "technology of citizenship"—part of the material and immaterial mechanisms of everyday life that help make people into self-governing citizens.[19] Through a range of material and social sites—Enlightenment punishment, Victorian transportation, antebellum American commerce, Machine-Age manufacturing, and Cold-War global warfare—their efforts gave shape to ideas about breakdowns that persist into our own times. Mumford's lament about indifference was part of a long tradition of desire: a culturally and politically powerful aspiration that people should, in some way, be moved by the breakage of machines.

Despite the long neglect that Mumford and others initiated, a rich and growing literature has begun teaching us how crucially breakdowns matter to our world. "Faults are defaults," as Simon Schaffer puts it.[20] But beyond their ubiquity lies their historical significance. The historian On Barak explains that "accidents, breakdowns, and malfunctions are vital forces of technological advancement." Train crashes shaped the development of the railway, "with its technological safety measures, synchronization schemes, and operation protocols." The shipwreck helped give modern sea vessels their material form.[21] Beyond shaping technologies, breakdowns structure our very categories for thinking about the technological, social, and natural worlds. In the form of blackouts or broken levées, they define the limits and agencies of nature.[22] They shape late-modern networks of disposal and discard that help define the Anthropocene.[23] And in the complex technological systems of our own time, they morph and combine through dizzying and improbable chains of cause and effect to make accidents "normal," both shaping and explaining the risk society and its cultures of disaster.[24]

Despite those insights, though, we still have a very poor understanding of breakdowns as historical objects. Some scholars have attempted to theorize breakdowns by turning to "new" materialisms that cast matter as "vibrant" and "alive" and breakdowns as a recalcitrant and obdurate agency of machines.[25] However, as Taylor Moore has observed, that view continues to separate the actions of nonhuman objects from the broader social and cultural histories of the humans that live with them.[26]

Breakdowns, for instance, did not always exist. Of course, machines have always found ways to break. But what we think of as a breakdown—a pathological interruption in the operations of a machine—is a relatively recent historical phenomenon. Before the eighteenth century, European-based languages generally lacked a word for these occurrences. Early modern commentators relied on a more general language of mischief, destruction, injury, and damage that seemed to endlessly talk around the events we casually describe with a single word.[27] Beginning in the eighteenth century, a number of developments—the rise of a passive strain of mechanical philosophy, growing secularization, changed understandings of chance and causation, an emerging public sphere, the rise of wage labor and contract law, to name a few—created conditions for breakdowns to emerge as a modern phenomenon that we might recognize today.[28]

A number of features characterized this emergence. Conceptually, the breakdown drew on Enlightenment visions of machine perfectibility that made it appear exceptional rather than routine. Temporally, it occupied a moment between a past in which machines had worked as expected and a future in which they would work that way again. Spatially and materially, it developed in close relation to automatic machinery. As Roger Cooter and Bill Luckin explain in the analogous case of accidents, the places where things were most predictable were also the fertile grounds where mishaps could emerge most conspicuously.[29] Finally, the breakdown was defined by its *overdetermination*: A breakdown could always be causally accounted for in more ways than were strictly necessary, always vacillating between the human and the material. As Peter Galison illustrates in his examination of aircraft disasters, forensic crash investigators might track an exploding engine to a flaw in a titanium fan disk or to the maintenance technician who failed to notice it.[30] That oscillation between human and material faults runs through all the histories I explore in this book and persists into our day. Giving a "modern" explanation for a breakdown has therefore meant describing an event where, precisely because humans or materials could always be pressed into the service of blame or explanation, the qualities of people and materials could never be completely separated.

This intertwining of the human and the machinic is at the center of *The Broken Machine*. And it is part of what makes any history of breakdowns also a history of the self. In his classic study of self-fashioning in Renaissance Europe, Stephen Greenblatt defines the self as "a sense of personal order, a characteristic mode of address to the world, . . . a consistent mode of perceiving and behaving." These ways of being change over historical time but, Greenblatt argues, they are consistently enmeshed in efforts to shape an identity either for oneself or for others and often in alignment with the demeanor of the elite.[31] Although Greenblatt highlights the role of literature in that process, technologies writ large have helped shape historical selves across millennia. Although historians of technology have generally avoided the self as a topic, cultural historians and philosophers more broadly have embraced the role of technologies in shaping historical ways of being.[32]

That literature has generally understood the self as a project or artifact realized through the material operations of *working* technologies. Drawing on Foucault's expansive understanding of "technologies of the self," for instance, Lorraine Daston and Peter Galison have demonstrated how lab notebooks, grid drawings, and scientific atlases were crucial material supports for the objective scientific self of the nineteenth century.[33] As Bruno Latour has additionally explained, getting machines to work in the first place requires us to be certain kinds of people, reinforcing how working technologies shape historical "modes of perceiving and behaving."[34] Feminist scholars have pointed to the exclusions created through those relations, as well as the role of technologies in sustaining, preserving, and affording new possibilities of selfhood.[35] Here, too they have focused their attention on functioning technologies. As Judy Wajcman expresses it, both masculinity and femininity "acquire their meaning and character through their enrolment and embeddedness in *working* machines."[36] Similarly, a rich and powerful scholarship demonstrates how race has been made, sustained, and protected through working devices. Alondra Nelson, Ruha Benjamin, and Safiya Noble, to name only a few, have revealed how our current digital technologies perpetuate and amplify social hierarchies, reify racial categories, and encode discrimination, not as incidental or accidental byproducts, but rather as part of the very definition of what it means for these technologies to "work."[37]

This rich literature has been crucial for helping us think through the relationship between technology and selfhood. But if we take the emphasis on working technologies too far, it can also mislead into thinking that the relationship between machines and selves ends when things break. Instead, I understand broken machines as objects around which the desirable and proscribed properties of people also take historical shape, and by design. It may be tempting to think, as Sara Ahmed suggests, that broken things are of no use.[38] But broken machines have figured centrally in projects to define people over the last three hundred years. They were deliberately woven into "characteristic mode of address" in the periods and contexts I focus on in this book. Scholars of media and technology, and of the growing interdisciplinary field of failure studies, have emphasized this powerful role of breakdowns and the self in our contemporary world.[39] But those current relations are themselves a product of the long and rich histories of breakdown and self-making, including the ones I explore in this book.

Understanding this historical role of breakdowns helps us make sense of the full range of ways that technologies have been used in the difference-making projects of modernity.[40] In particular, breakdowns have functioned as instruments for drawing what Ann Stoler calls "the moral parameters of European nations," delineating the modern and the backward, the respectable and the rabble, the citizen and the "other," both inside and outside Europe's borders.[41] In the 1920s, in one of the earliest sociologies of breakdowns we have, the French-born German economist and philosopher Alfred Sohn-Rethel used breakdowns to place Neapolitan society outside modern European time, for example, casting the Neapolitan acceptance of broken machines as a quaint throwback to the seventeenth century. "In Naples," he explained, "technical devices are, as a rule, broken: it is only under exceptional circumstances and due to some astonishing accident that something will be found to be intact. . . . Mechanisms cannot, in this city, function as civilization's continuum, the role for which they are predestined: Naples turns everything on its head."[42]

That use of broken machines as measures of modernity extended far beyond Europe. As Michael Adas famously noted, early in the modern period, European travelers outside the West cast their technologies as "absent objects" that defined the inferiority, pre-modernity, and deficiency of non-Western societies.[43] As those technologies spread, though—as steam engines, telegraphs, and electrical networks increasingly occupied the Antilles or the Horn of Africa or the Indian subcontinent—at least two things happened. First, merely possessing those artifacts no longer served as a marker of Western superiority; and second, the technologies themselves revealed their tragic vulnerabilities in foreign environments.[44] They broke down. As Brian Larkin notes, this material betrayal of artifacts was of a piece with the betrayal of elite promises and the vulnerabilities of Western empire.[45] As Larkin explains, "perhaps nothing reduces the grandeur and majesty of technological achievement more than breakdown and failure. The plane cannot take off, the bridge that sways cannot be crossed, the distorted radio signal all evoke laughter or frustration, not fear and awe."[46] But precisely in these moments, the attitude *toward* breakdowns came to function as a mark of difference and superiority. The places where machines were always breaking, and the tolerance for that breakage—the inability or unwillingness to be troubled by breakdowns—represented an inferior way of being that the West had ostensibly left behind.[47] This was why Mumford could claim (and lament) its demise. Breakdowns now redrew a map of modernity. The rest of the world was a world in disrepair, a "broken world." It is only by recognizing how broken machines served, in Achille Mbembe's phrasing, as polemical arguments "for the West's desperate desire to assert its difference from the rest of the world" that we understand the full scope of the ways that modern technologies have been used to define and divide human populations in ways that continue to this day.[48]

How to Write a Disturbing History

Writing a history of breakdowns involves grappling—socially, culturally, philosophically—with the kinds of objects they represent. For guidance, it is tempting to turn to the influential work of the philosopher Martin Heidegger.[49] Heidegger was a contemporary of Mumford. *Technics and Civilization* was going to press just as Heidegger was pledging his loyalty to the Nazi Party.[50] His antisemitism and his support of National Socialism have made many scholars rightly wary of adopting his perspectives uncritically. Tiago Saraiva, for instance, has brilliantly analyzed the unsettling historical contexts that produced Heidegger's now-popular concept of "things."[51] Saraiva warns us against indiscriminately invoking that term to describe all historical objects—from fascist agricultural products to the scientific data of democratic debates—as if they gather natural and social worlds in the same ways.[52]

Breakdowns deserve similar scrutiny. Heidegger used breakdowns as a transhistorical and disposable philosophical object—a catalyst for revelation that he discarded once it furnished its powerful insights. Science and technology studies scholars have used breakdowns, and discarded them, in precisely this way when examining everything from airplane crashes to failing infrastructures.[53] But, like "things," Heidegger's breakdowns had a dark history. Conjured in a hut in the Black Forest in the 1920s, their premodern, mystical quality abetted an escape from the modern industrializing landscapes and liberal politics of the short-lived Weimar Republic, both of which Heidegger opposed. Invoking a mythical German past rooted in romantic nationalism, its racist conceptions of *volk* and social order, breakdowns resonated with the violent and reactionary politics that would soon guide the National Socialist movement. And they mobilized an element that has gone largely unremarked by the historians who draw on them, an element that was central to their power for Heidegger—namely, their relationship to the self. Any broken tool, Heidegger explained, "is *disturbing* to us."[54]

One of the premises of *The Broken Machine* is that treating breakdowns as moments of analytical clarity, as Heidegger did, has obscured their deeply historical politics of the self. Delving critically into Heidegger's casual mention of breakdowns as disturbances is one way of recapturing that history and is critical to the story I want to tell.

As an assault on peace, tranquility, and regularity, "disturbance" calls to mind exactly the transitory and ephemeral properties that have made breakdowns into objects outside of historical time. But two additional understandings of disturbance are telling here. The first is its psychological dimension. Within a century of entering English in the late Middle Ages, a disturbance had become an interruption or perturbation of mental tranquility.[55] Heidegger's usage suggested that, more than a temporary disruption in machinic states and human activity, breakdowns were disturbances that troubled (or ought to trouble) the self. The second meaning, though, may be slightly less obvious. It comes to us from ecology and is helpful for understanding how we might make sense of these events historically. In ecological systems, disturbances are also discrete disruptions of ordered worlds—perturbations within an ecosystem or animal population, for instance. Floods, volcanic eruptions, storms, grazing animals, even deforestation. But they are not *only* disruptions. By altering environments, even slightly, disturbances create new possibilities and renew relations that contribute to the long-term stability of the whole, releasing nutrients from the soil, for instance, nurturing interspecies relations, or sustaining necessary cycles of regeneration.[56] As Anna Tsing notes in *The Mushroom at the End of the World*, the matsutake mushroom of Japan cannot grow without a partner: the red pine. But red pines cannot successfully compete against faster-growing broadleaf trees, particularly oaks, without human disturbances in the forests where they germinate. As Tsing explains, "disturbance realigns possibilities of transformative encounter." In doing so, it challenges assumptions of progress and development that we build into our accounts of the world, including our understandings of nature, technology, and society.[57]

Seen as interruptions, pathologies, and ruptures, breakdowns are temporary rips in the social fabric, mended through maintenance and repair. Seen as disturbances, however, breakdowns become elements of dynamic social order in a way Heidegger possibly recognized but concealed.[58] Elsewhere, I have written about how, over the time span covered in this book, failing machines helped define the qualities and agency of the natural world, supporting categories of the "marvelous" and the "hostile" that were central to projects from imperial expansion to planetary-scale warfare.[59] A similar technopolitics permeated understandings of social order.[60] Historically, people have used breakdowns as moments to privilege, cultivate, and assert certain ways of being while denigrating others. This technopolitics, intertwining failing machines with kinds of people, has continued to invest people in both the workings of machines and the maintenance of the social order, even during moments when those technologies were not working at all. Rather than *suspensions* of the social order, then, breakdowns have formed *extensions* of it. To rephrase Latour, even broken technologies are society made durable.

Rather than move quickly past breakdowns, treating them as either annoyances or transhistorical objects in the service of philosophical insights, then, what would it mean to stay with these events and give them a proper history? To let the smoke clear and the dust settle and see what they can teach us about the worlds we thought we knew? The most speculative argument I put forward in this book is that the historical intertwining of breakdowns and the self persists into our own time. Jeffrey Schnapp has written about "the density that characterizes the object world: the subtle incrustations of intention and invention, fantasy and ideology, tradition and accident that . . . an object carried in the train of its existence."[61] New machines don't simply replace old ones in a monarchical succession. The material world accretes.[62] Printed books coexist with coal-fired power stations and artificial intelligence. Shaped in relation to those technologies, historical selves accrete with them.[63] In that building up is a statement about the kinds of selves that have been replicated and reproduced and embodied in technologies and the kinds that have been discarded and pushed to the margins. Recovering that process means working against the modern ambition to turn moments of disturbance into moments of opportunity—and then to promptly forget them.[64] In that amnesia lies the possibility of another history.

Breaking it Down: The Structure of the Book

Given the way breakdowns fill the modern period, it would be impossible to give an account of them in all their richness. *The Broken Machine* explores histories of breakdown that unfolded in the modern Atlantic World. For three decades now, that site has been a focus of histories that helped define not only what modernity was and how it related to machines but also who belonged to it.[65] As a result, it is a place where the questions I want to explore are represented most clearly. Historians have debated the boundaries and nature of modernity for decades, bordering on centuries. I focus on what is sometimes called classical and late modernity—the period from the French and Haitian Revolutions in the late eighteenth century to the decolonization movements of the Cold War in the second half of the twentieth century.[66] I use this framing because I understand modernity to involve a set of societal transformations beginning in the eighteenth-century Atlantic World that created what we could think of as the modern condition. Beginning with revolutionary France, I track disturbing histories of breakdown to Victorian Britain and then to nineteenth- and twentieth-century Anglo-America to see how authorities, experts, and political elites cast them as sites that wove together governance, the self, and the material failings of modern machines.

Rather than trace a continuous historical thread, each chapter examines a specific historical moment, a specific technology, the politics of breakdown it supported, and an understanding of the kinds of people it helped define. Focusing on these well-studied contexts and periods isn't an attempt to deny the importance of other places and times or to reproduce what Tina Campt has called the "sovereign gaze" of the regimes that generated these ways of thinking.[67] Just the opposite. It is precisely because administrators and elites within these societies believed their attitude toward breakdowns made them different, modern, and superior that they provide a chance to confront these powerful ways of thinking with their own history. And it is precisely because breakdowns still trace out global geographies of difference that we can use these particular places and times to historicize the experience of a transatlantic world where caring about breakdowns, according to Mumford, disappeared long ago.

The book begins with an episode we have already sketched, one drawn from the histories of imagination, sentiment, and political turmoil. While French revolutionaries debated the abolition of slavery to temper a revolution in Haiti, and while enslavers cast Haitian revolutionaries as unruly machines, the French state entrusted its capital punishments to a new device: the guillotine. We think we know why, but the chapter turns the conventional history of that machine on its head. Focusing on the crucial period before the Reign of Terror, I argue that rather than a simple embodiment of Enlightenment rationality, the guillotine represented a "sentimental machine"—a device whose very form embodied the virtues of eighteenth-century French sentimentalism and of the contemporary mechanical arts, including its ideas about breakdown. A citizen disturbed by the effects of a breaking guillotine was a citizen displaying the *sensibilité* that underpinned the new nation, its educational reforms, its festivals, and its core virtues. When the actual breakdown of the guillotine betrayed that expectation, authorities used the event to condemn the character of its builder, Tobias Schmidt, and to draw the boundaries of the social order to push him to the margins.

Chapter 2 moves across the English Channel, where visions of British social order developed in the long shadow of France's revolution. Here, I focus on questions that middle-class Victorians asked about the breaking machines that increasingly surrounded them and specifically about what breakdowns said about their society and themselves. I explore these questions by examining how material breakages in the "machinery of railways," and the spectacular train accidents they produced, formed more than interruptions to Victorian life, circulation, and commerce. They formed places where Victorian experts worked out questions of culpability and causation and, with them, conceptions of self-responsibility, self-regulation, and self-governance that underpinned liberal visions of Victorian society. The military-trained engineers charged with investigating breakdowns in the context of calamity developed a new kind of accident—the "system accident"—as a way of resolving uncomfortable questions about the Victorian self and culpability: who should be troubled by the breakdowns of industrial modernity and how much. Through that history, I chart how systems went from abstract schemes of work to the material assemblages we understand today.

Chapter 3 forms a "break narrative" that pauses to explore an alternative history of breakdowns in antebellum America. Before the "breakdown" was either a mechanical event or a psychic crisis, it was a celebratory dance performed by enslaved Americans during a pause in harvest labor. The chapter examines how these origins intersected with the experiences, mythologies, and perils of machine breakage in the antebellum United States, linking the subversive activity of dance and slave celebration to labor suspensions, material disruption, Black identity, and the possibility of self-emancipation. Tracking these connections across bodily performance, steamship boiler explosions, and the "malfunctions" of the cotton gin, the chapter challenges recent scholarship by seeing antebellum "breakdowns" as a site for the contested expression of a Black technological self that carried through the Gilded Age and Progressive Era—the subject of chapter 4.

Chapter 3 deserves a special word. It represents a particular kind of disturbance in our histories of technology and has three specific aims. The first is to shift our perspective on the histories that lead up to, and away from, it. It is meant to disorient us after the book's opening chapters. By providing vignettes of breakdowns in Black America and placing them at the book's core, it highlights how the chapters that have come before have been about constructions of race and technology that we take for granted. And it ensures that we cannot see the histories that follow in the same way once we've left the chapter behind. Second, in its form and length, it is meant to underline the particular problem of sources in writing Black technological history during this period. We should feel disturbed in reading it and question our dissatisfaction with the fact that we don't (and sometimes can't) know more. Lastly, in its content, the chapter is meant to propose a different set of understandings of breakdown and the self—a counter-history to the arc of the book.

Chapter 4 explicitly engages the question of how breakdowns drew lines of race and technology. It examines how one of the most influential graphic technologies of scientific management—the Gantt chart—grew out of attempts to create intricate psychic and cultural linkages between two kinds of failure in Progressive Era America: failure as a condition of industrial machinery and failure as a kind of person. The "paper technologies" of scientific management—charts, diagrams, instruction cards, and questionnaires—were not only devices for Machine Age Americans to account for their labor, their moral character, and their time; they were also instruments through which breakdowns were directly tied to the interior lives of factory workers, specifically machine operators, as a way of transforming these events into a problem of the white laboring self. For Henry Gantt, the creator of the eponymous chart, the graphs formed part of a project of racial uplift that linked white manhood to industrial machines, and a parallel project of racial containment that kept Black workers out of northern factories during the Great Migration. Tracing those concerns about failing white men and failing industrial machines across steel mills, metals shops, and textile factories, the chapter illustrates how efficiency experts, factory managers, and white workers came to understand failing machines as a problem of the kinds of people who belonged in the theaters of American industry and their role in the civilizational mission of turn-of-the-century America.

Chapter 5 continues the emphasis on morality and machines, investigating how officials in Cold War North America saw electronic breakdown as more than a straightforward problem of materials or machines. They understood it as a crisis rooted in the problematic actions, capabilities, and attitudes of two groups: the circuit designers tasked with developing electronics and the technicians charged with maintaining them. In response, military officials mounted a moral crusade exhorting military personnel to identify with the failures of their technologies; and they devised practical programs that used breakdowns to shape and define values of the reasonable and rational thinking self of the Cold War. Focusing on these two sets of technical workers, the chapter explores how making electronics reliable became a problem of making technicians think in reliable, trustworthy ways, cultivating a self that mirrored and extended the ideal citizens of Western liberal democracies. The techniques developed to guide and manage the minds of designers and technicians not only echoed Cold War social scientific theories about how to build strong democratic institutions and ensure the reliable behavior of citizens, but also shaped the material culture of electronics in the late twentieth century, embedding this idea of the self in its very circuitry.

On its own, each chapter is designed as a stand-alone intervention. It takes a well-known episode—an influential history of technology that we thought we knew—and rewrites it through the disturbance of the breakdown. The guillotine turns out to be a different object than we imagined: sentimental rather than rational. The Victorian railway "system" turns out to be not chiefly a complex material assemblage but rather a novel way of parsing the responsibilities of fallible Victorians and faulty machines. The charts of American scientific management embody not an ideal of maximal efficiency but rather a profoundly racialized anxiety over "failure" and "idleness" as both a breakdown of industrial machines and as a kind of person. And the electronic failures of the Cold War signal the moral failings and rational fallibility of the Western democratic citizens charged with designing and maintaining them.

The primary significance of these episodes, however, lies in the work they do collectively. First, by reinterpreting these well-studied episodes, *The Broken Machine* suggests a different understanding of the history of modern technology, where questions of social order, power, and the self were worked out around failing machines as well as functioning ones. Second, the chapters iteratively illustrate, across two centuries and two continents of the modern North Atlantic world, the book's main argument: that breakdowns were not interruptions in the formation of modern selves but rather extensions of it. And finally, the choice and arrangement of the episodes traces a specific succession of historical selves—sentimental, responsible, efficient, reasoning—whose enormously influential legacies have shaped our contemporary politics of breaking machines. Taken as a whole, they demonstrate that if we want to understand the full relationship between modern technology, the self, and social order, we need to see breakdowns as more than passing nuisances or perishable instruments of a more enduring clarity. These specific historical episodes and their legacies serve as a crucial guide.

A word about terminology. I try to avoid any essentialized and transhistorical understanding of breakdowns, instead preferring a historical inquiry into how the unanticipated, unexpected, and (from a certain viewpoint) undesired technological behaviors of the time were enrolled in political projects of meaning, inclusion, and difference. I use the term "breakdown" here mainly as an analytical term to describe a range of material behaviors of machines that do not necessarily involve complete stoppage. You will hear others—malfunctions, failures, short-circuits—and I treat them as elements of a larger group of phenomena that we might think of as "breakdowns."[68] The one place where this term most closely aligns with an actors' category is chapter 3, where I want to point out the connections and resonances between the breakdown as cultural form and its use to denote a mechanical failure. Recapturing those understandings revives the richness and specificity of the term while making clear the cultural work that has shaped our own understanding of these events and given them coherence as a category.

A word, too, about what this book is *not* about. Although it focuses on the self, it is not primarily a history of emotions, affect, or interior life. Rather, it is a history of how various powerful groups imagined the inner lives of people and their relationship to breaking machines. Like scientific controversies, the rhythms and convergences of breakage produce moments where people openly articulate some of their most taken-for-granted assumptions.[69] Rather than trying to look directly into the self, we can use these moments to explore how authorities and elites imagined one of the formative relationships of modernity, how they tried to align the inner lives of others with their own visions, and how the people at the center of their efforts embraced, evaded, or rejected them.[70] Second, although the book traces ideas about why machines break, it is not an intellectual history of the modern material sciences. There are many fascinating histories that chart the processes, practices, and theories leading to ever more reliable machines. This is not one of them. The book, then, is not a history of feeling itself or a history of engineering per se. If anything, it is a history of the repeated and converging attempts—often misguided, frequently unsuccessful, always fascinating—to engineer what we should feel in the face of breaking machines.

This book, then, hopes to be a disturbance of its own. It aims to both interrupt and trouble our taken-for-granted ideas about technology and its role in our lives. It begins from the premise that the stories we tell about technology matter and that our current historical moment might need new ones. Breakdowns *are* portals into worlds we thought we knew—worlds about which we imagined not much more could be said. But, against Heidegger, they are portals into the cultural and political histories of their own times. Moments of breakage reveal not only how our machines have shaped us, or how we have historically built ourselves into our machines, but also how our technologies have defined insiders and outsiders, patriots and traitors, good citizens and bad, even as they failed. If our current moment takes working machines for granted, this book is an attempt to confront that tendency with its own history. Here, then, is a beginning.

Figure 1.1

Cour de Commerce St. André des Arts, 1866. The street's southern end (in the distance) was shortened in the 1870s. It is otherwise largely unchanged since the 1790s. Café Procope is the low building on the right. Marat's printing press was near the end of the street on the left. The guillotine was first built in a shed (now gone) below the open second-floor windows on the right-hand side. *Source:* Charles Marville. Bibliothèque Historique de la Ville de Paris. Fonds Charles Marville—Cote: 2-EPR-00359, CARPH000807.

1 The Unfailing Machine: A Sentimental History of the Guillotine

All social questions achieve their finality around that blade.
—Victor Hugo, *Les Misérables*

April 1792, Paris. The former Cordeliers district. It would have been easy to miss the narrow cobblestone laneway in the tangle of buildings on the Left Bank. Before 1776, the small shop-lined street did not even exist. Older maps of the city showed the medieval and early modern ruins it had been carved from—a fortress wall, a moat, handball courts. Before Baron Haussmann's restructuring of Paris in the 1870s amputated its southern extremity, the street was only 160 meters long. At each end, its arched entryways ducked beneath apartment buildings before spilling onto the bustle of Rue Saint-André-des-Arts or Rue de L'École-de-Médecine. But this was no shortcut on the way to somewhere else. You had to *want* to find yourself here—for the booksellers, the grocers, the artisans; for the laundress Lemor, the engraver Gateaux, the printers Brune and Duplain, the candlemaker Bourjot, or the artisan Bernard and his ingenious medical devices. By the spring of 1792, there had been at least one flirtation here. And one murder.[1] The surrounding warren of streets and alleys formed one of the most radical quarters of revolutionary Paris. Camille Desmoulins, who lived for a time at the street's southern entrance, called the area "Little Sparta" and claimed that he knew all its residents by sight.[2] So many ardent revolutionaries called it home that it drew comparisons to the Temple of Vesta, whose sacred fire foretold the fate of the Roman Republic. Most of the residents, though, simply knew it as the former Cordeliers district.

In the middle of this radical body politic, the small street known as the Cour du Commerce formed a vital, pulsing artery. By day, it was a node in the underground network of grub street literature that spilled from Duplain's printing press and helped fuel the Revolution.[3] At night, it provided a shadowy back entrance to Café Procope, where Voltaire and Rousseau feted and brooded over their plays, where Diderot and D'Alembert launched the *Encyclopédie*, and where Jacobins and Cordeliers held court alongside *comédiens* and gossipmongers. Everyone you needed to launch a revolution was here.[4] The Revolution's "savior," Danton, would soon be arrested in his apartment above the street's southern entrance. The period's most celebrated and feared publisher, Marat, would be murdered only a stone's throw away. But those events were yet to come. For now, in April 1792, in a shed belonging to the laundress Lemor, across the street from Brune's printing press and next door to Procope, a German émigré piano maker named Tobias Schmidt was performing a series of bizarre experiments with his latest mechanical creation—an "unfailing machine" that, in less than two weeks, would become the emblem of the Revolution.

Schmidt was no stranger to unusual inventions or to controversy. Residents of the Cordeliers knew him as a creator of ingenious machines and musical instruments advertised in the *Journal de Paris*. The Revolution had only amplified his energies and his creativity. In early July 1790, he became embroiled in a complicated priority dispute over a "hydraulic diving machine" that he demonstrated by plunging into the river Seine and remaining submerged for half an hour. His exact movements in the eighteen months that followed are something of a mystery. In January 1791, he was still on the Cour du Commerce, appealing to the National Assembly to recognize his priority over the diving machine.[5] By February 1792, some accounts place him in Strasbourg, allegedly working for an enterprising notary named Laquiante.[6] But by April 1792, he was very obviously back in his workshop on the little street and hard at work, still a relatively obscure but hustling and ambitious artisan.

The machine taking shape in that shed had a complicated genealogy. Exact instructions for building it had been drawn up by one of France's luminaries—the perpetual secretary of the Academy of Surgery, Antoine Louis. He, in turn, had considered two versions of the device: the first, suggested verbally by Schmidt, and a second, inspired by a hypothetical machine that the physician and politician Joseph-Ignace Guillotin had imagined. (Guillotin's apartment was conveniently located within shouting distance of the laboring Schmidt.) Louis had ultimately opted for his own version of the machine, drawn from his vast knowledge of mechanical inventions, physiology, and death. He had painstakingly set out its material details: the dimensions of the timbers, the quality of steel, even the thickness of the rope. Running through his careful material specifications was one overriding imperative: *The machine had to be infallible*. And yet, within a few months, the device's breakdowns would place it at the center of both a priority dispute over its invention and a disputed vision about the relationship between failing machines and the selves that underpinned the French Revolution. As Schmidt's prototype took shape that April, few of the passersby could have realized how their lives would become entwined with it as it dispatched the neighborhood's most famous residents—Danton, Fabre D'Eglantine, Desmoulins—and more than a few of the regulars at Procope.

We have come to know Schmidt's device as the guillotine—a name that obscures its complicated genealogy. This chapter puts aside the question of priority and invention that even Schmidt and his contemporaries debated.[7] Instead, it asks another question that may seem too simple to pose: what did authorities during the French Revolution believe was at stake in the machine's failure? The guillotine, we have been taught, was a rational machine: born out of the Enlightenment drive to establish a uniform and humane method of execution for all French citizens, to eliminate criminal suffering, and to elevate rational civic virtue over sentiment.[8] Its failure, it seems, would have called all this into question. But this received history is rooted in the perception that, if anything, the guillotine worked *too* well. Its brutal efficacy underwrote successive and sometimes contradictory visions of revolutionary social order. Ironically, so the story goes, the guillotine's experiment in rationalization and enlightened punishment went horribly awry, as the machine fell into the hands of the Revolution's increasingly extreme elements—most famously, Robespierre and the Jacobins. All too soon, it became the "ensign" of the Revolution and the sinister instrument of its darkest phase: the Reign of Terror.[9] In that transformation, the image of the guillotine's infallibility, soon morphing into a myth, was secured.

This chapter tells a very different story. It argues that the experience of criminal suffering was never the central concern of the men who dreamt, designed, and commissioned the guillotine. Instead, the machine was shaped—materially, politically, culturally—by concerns about the people *watching* the execution.[10] And those concerns were, from the very beginning, shaped by an expectation of what people should feel if the machine failed. The fear of machine failure was, first and foremost, a fear about the effects of a failed execution on the sentimental selves that legislators and revolutionary authorities imagined filling the volatile public spaces of the early Revolution and underpinning its visions of social order. This imbued the guillotine with a specific politics of breakdown: one where proper citizens would be disturbed by even the idea of a failing machine, precisely because its horrific consequences echoed the horrors of the Ancien Régime. This made the device's functioning part of what William Reddy has called the "emotional regime" of the period—the "normative order for emotions," expressed through official rituals and practices, that underpins the stability of political regimes.[11] Rather than ignoring the guillotine's effect on public emotion, as Daniel Arasse has claimed, its bourgeois proponents and inventors did precisely the opposite. They designed what they believed was a mechanical safeguard against one of the greatest threats to social order: the public dangers of sensing and feeling too intensely.[12] When none of this went to plan—when the breakdown of the guillotine left spectators unmoved—the Revolution's political class used those events to set themselves apart from its insensitive rabble, including the artisan Schmidt. They used the reaction to the broken machine as a way of defining who belonged to the Revolution's sentimental social order who stood outside it.

Focusing on what the breakdown of the guillotine meant in the summer of 1792 can transform our understanding of its history as a revolutionary technology. To see the machine in this new light means repositioning it historically in three ways. First, we need to locate the device firmly in the period before the Jacobin Reign of Terror (1793-1794), when the machine was still a viable instrument for a reconstructed and reformed social order—a "theatre of proof" for the Revolution's loftiest ideals.[13] Second, we have to position the device not only within the history of rationality, to which it certainly belongs, but within a history of sentiment. Eighteenth-century French sentimentalism deeply shaped not only the revolutionary authorities' ideas about the role of emotion in creating new French citizens, but their understanding of how sentimental virtues could be reflected or threatened in the form and functioning of its machines.[14] Finally, we will need to mount the creaking scaffold, as unsettling as it might be, and look beyond the spectacle and pageants surrounding the criminal body, turning instead to face the body public, the theories of self used to make sense of it, and the role of broken machines in defining it. If we do that, then the guillotine emerges as a very different kind of device: a *sentimental* machine born out of the temporary union between a profoundly public, late eighteenth-century French sentimentalism, on one hand, and the contemporary mechanical arts, with their resonant theories of breakdown, on the other.[15]

Sentimental Machines

Schmidt's machine may have been built in a rough shed on the narrow and bustling Cour du Commerce, but its philosophical origins lay in the genteel salons on the other side of the river Seine. A mixed space of men and women, nobles and bourgeois, these theaters of conversation and sociability were initially extensions of court society run by powerful *salonnières*—women who selected the guests and topics and led their patrons in cultivating the salons' central virtues: *politesse, civilité,* and *honnêteté.*[16] Starting in the 1750s, these spaces became seedbeds for a theory of mind that would become known as "sentimentalism."[17]

Sentimentalism drew heavily on French interpretations of the English philosopher and physician John Locke, first by the Baron de Condillac and later by Denis Diderot, who seized on Locke's sense-based theory of knowledge in the *Essay Concerning Human Understanding*. Locke's theory saw the mind as a blank slate at birth and the senses as instruments that inscribed the outside world on it.[18] Objects "entered" the mind by causing changes in the sense organs, which were experienced as sensations. These sensations, in turn, gave rise to perceptions and to the higher faculties—memory, imagination, abstraction, judgment, reason. All sentiments, on this understanding, had their origins in physical sensation; and those same sentiments, properly managed and cultivated, were as important as reason in underpinning knowledge, morality, and social order.[19]

Sentimentalism swept the elite social geography of pre-revolutionary Paris from the 1760s onward and quickly spread beyond the culture of the salons and the *hôtels particuliers* to the wider literary and popular culture of the late Ancien Régime. The popular "theater of tears" or novels such as Rousseau's *Émile*, the best-selling novel of the eighteenth century and a later influence on the Jacobins, modeled and celebrated the act of being moved.[20] Voltaire saw himself as a "man of feeling" and mounted his attacks against religious intolerance on that basis.[21] As a theory of the self, sentimentalism explained not only the sensory origins of sentiment, and not only how people should cultivate themselves as persons and subjects, but also how sentiment could spread to form a common feeling among strangers and the basis of a social bond.[22]

Like so many of the Enlightenment's universalisms, sentimentalism was also an instrument of social difference. It formed what Dorinda Outram has called an "intra-class" language, that legitimated many of the privileges, attributes, capacities, and prejudices of the bourgeois circles that cultivated it.[23] Although women commanded the physical and social spaces of Parisian sentimentalism, the movement's principal architects were wealthy white men. Their theories cast women as delicate and susceptible to rapidly changing sensations, making them voluble, unqualified for public education, and uninterested in politics or science (two spheres where male sentimentalists were particularly prominent).[24] The movement supposedly disdained wealth but held that proper sentiment was a sign that the wealthy deserved their privilege.[25] More generally, the sentimentalists used feeling to determine social value, dividing the world into the mostly bourgeois selves who sympathized, on one side, and the objects that inspired their sympathy, on the other.[26] Their theory emoted empathy for the poor but largely placed "the people" outside its full potential for knowledge and morality—tainting them with a dangerous insensibility that made them susceptible to imaginings, charlatanry, fanaticism, and rioting.[27] Sentimentalism flourished at a time of colonial expansion and chattel slavery. Its powerful rhetoric turned the enslaved Black peoples of the French colonies into objects of pity and supplicants for emancipation. But although it stoked calls for abolition, it also cast those same people as objects rather than agents in its theories of humanity and the common good, helping to underpin the Enlightenment's racial contract. Ultimately, sentimentalism placed more emphasis on the common feeling that united like-minded abolitionists to one another than on what linked them to the enslaved people whose cause they defended.[28]

Although it dealt with the rarified air of emotion, sentimentalism formed a material movement as much as a philosophical and a social one. Its literature often featured tales told by objects—snuffboxes, slippers, and pocket watches that narrated their travels and abuse at the hands of unworthy owners before finding a person of virtue.[29] By the 1780s, the links between sentiments and material culture were a fixture of the medical profession (Locke himself had been a physician), informing how surgeons and doctors thought about the body and the machines they used to intervene in its operations.[30] Sentimentalist devices filled pre-revolutionary Paris, where they often served as sensory aids. Valentin Haüy, founder of France's premier school of the blind, where Louis Braille would eventually study, advertised a printing press that printed letters in relief but without any color.[31] French journals praised the invention for making mental objects sensual and material, and they hailed Haüy as "having found the means to render ideas tangible." On the Cour du Commerce, the resident instrument maker, Bernard, advertised a "hearing instrument to combat deafness" that could easily be hidden under the hair.[32] In folk remedies, popular fashion, and surgical and medical instruments designed by local *artistes* and artisans, sentimentalism inspired material conduits into inner lives.[33]

Sentimentalism's greatest influence on the technologies of the late eighteenth century came through its emphasis on "simple machines," part of a contemporary emphasis that drew on sentimentalism's distaste for concealment, deception, and duplicity.[34] Machines have historically been associated with trickery, with "machinations," because their workings are often hidden and seemingly detached from their effects.[35] For the sentimentalists, not only did complex machines magnify obfuscation, they were also more likely to produce *inconvénients* or breakdowns. Both the opaque operations and the more frequent breakdowns of complex machines generally tracked back to the same moral shortcomings: the dishonesty of the artisans who built them.

Over the hundred and fifty years leading up to the French Revolution, ideas of mechanical simplicity countered this dark vision of machines, promising to make virtuous both the devices and the people associated with them. Simple machines, like simple humans, were honest and reliable. French commentators celebrated Benjamin Franklin, for instance, by explaining, "Franklin had genius, but still he had virtues, he was simple, good, modest above all."[36] His invention, the lightning rod, was a material embodiment of that virtue. Praised by artisans and academicians, resonating with allusions to ancient Greek mathematics and to Roman engineering, "simple machines" embodied straightforward construction, ease of use, and, not least, trouble-free operation.[37] Like the balance scale, the symbol of justice, their workings were transparent and verifiable, avoiding the possibilities of fraud and deception, including by their human operators.[38]

By the mid-eighteenth century, mechanical "simplicity" was powerful enough to be pitted against the virtue of "perfection"—the highest aim of the Enlightened mechanical arts.[39] As the century's most celebrated automaton maker, Jacques de Vaucanson, explained in the *Mercure de France* in 1751, "Sometimes, and this is completely normal, in perfecting a machine in certain respects, one complicates it in many others and multiplies its effects often at the expense of its simplicity."[40] Throughout the middle of the eighteenth century, the intertwining of mechanical simplicity and Enlightenment morality repeatedly linked the virtue of simple machines to the virtues of the self, nature, and society; and it connected human vice with complication, immorality, and breakdown.[41]

In the lead-up to the Revolution, this technopolitics of breakdown acquired political force as a metaphorical diagnosis for the sickness of the Ancien Régime itself. Rousseau used the complications of the "political machine" as a shorthand for the bloated and clunky state apparatus that would contribute to the Revolution.[42] Claude Adrien Helvétius (one of the period's leading sentimentalists), writing to Montesquieu, rejected the "complicated machine" of the constitution and instead envisioned a simpler mechanism of governance: "It will be a simple machine, whose springs, easy to direct, won't require that grand apparatus of wheels and counterweights so difficult to wind [*remonter*] for clumsy people, who are so often mixed up in governing."[43] Political commentaries exploited these metaphorical connections between mechanical complication and material failure. The *Gazette du Commerce*, commenting on the system of state finances in 1775, explained: "It is not easy [*aisé*] to reduce a strangely complicated and contradictory machine to the simplicity which assures operation, regularity, and endurance."[44] Those connections would carry over into the Revolution itself. In the tumultuous year of 1793, the increasingly radical politician Le Pelletier de Saint-Fargeau observed: "That which should serve as a foundation for all governments, and above all for the democratic commitment, is simplicity. This will be the case with a government that is too *recherché*, too delicate, too ingeniously [*savement*] complicated, like those ingenious machines that please, initially seducing on the account of their inventor, persuade even in their smallest effects, but that come to fail in their overall organization."[45]

By the summer of 1789, sentimentalism's combined social, political, and material emphasis formed a profoundly public psychology—outside the exclusive province of experts, spreading through salon culture and the vernacular of the literate classes and informing the political, moral, and material virtues of the early Revolution. Leading revolutionaries wove it into their political and cultural projects as a powerful instrument for shaping France's new revolutionary citizens even as they dismantled other elements of the Old Order.

As early as 1784, the Comte de Mirabeau, one of the Revolution's future leaders, explained: "In his capacity as a sensitive being, man is led much less by rigorous principles, which require thought to be seized in all their aspects, than by majestic objects, striking images, great spectacles, deep emotions. . . . Man, I will say again, obeys his impressions rather than his reason."[46] Mirabeau, who had a model of a slave ship built to emphasize the inhumanity of the trade, would place sentimentalism in the service of the Revolution, institutionalizing it in the schools and radical educational reforms of the period but also projecting its principles into the public theater of the capital.[47] The spectacles of the Revolution—the wearing of special clothing by public officials, the launching of tricolor balloons on the Champ de Mars, the cycle of festivals, the renaming of city streets, and the adoption of the new revolutionary calendar—would deploy sensation as a public tool through which citizens would become conscious collaborators in their own formation and where material objects, including machines, would play a critical role in creating the "omni-pedagogic urban environment" of revolutionary Paris.[48]

But Mirabeau's initiatives suggested that, as much as the sentimental self underpinned this new social vision, it was also a site of acute fear. As a privileged entry into the human interior, the senses had to be carefully regulated and managed. In a statement that would prefigure so many of the guillotine's multiple dimensions, Rousseau had explained: "What one too often sees, one no longer imagines, and it is only through imagination that we feel another's sorrows. By dint of seeing death and suffering [on a routine basis], priests and physicians become pitiless."[49] One particularly dangerous site for these effects was the burgeoning public spaces of the late Ancien Régime and the early Revolution, where sentiment could prove pathological, spreading through crowds and overwhelming the cherished virtue of reason.[50]

By the late eighteenth century, public executions represented exactly this kind of anti-sentimental spectacle and theatre. In the decades before the fall of the Bastille, a succession of brutal executions had horrified the Parisian public, beginning with the would-be regicide, Damiens, in 1757, whose four-hour torture and execution would be cited by legal reformers such as Beccaria and Thomas Paine and later used as a dramatic set piece by Michel Foucault.[51] More recently, crowds had thronged the scaffold and crowded the windows at the Place de Grève, only to see the botched beheading of the comte de Lally (figure 1.2), the disgraced governor-general of French India captured by the British and falsely accused of treason in France. Louis XVI would later exonerate Lally publicly and posthumously, but not before Voltaire and Thomas Carlyle denounced his execution as a judicial murder.

Precisely because these executions were increasingly seen as abhorrent and symbolic of the Ancien Régime, they occupied the intersection between sentimentalism's agoraphobia and its opportunism. There was danger in the brutality, cruelty, and suffering of these events. But the Revolution's sentimentalist leaders also recognized that a reformed execution could educate citizens in the virtues of the Revolution. Like the Revolution's festivals, the spectacle of the execution, with its open spaces and uninterrupted vistas

Figure 1.2
The botched execution of Lally, May 8, 1766, at the Place de Grève (now Place de L'Hôtel de Ville). Note the crowds of onlookers. The image illustrates the sentimentalist spectacles that executions represented and the danger they posed if things went wrong. The executioner, Charles-Henri Sanson, was also involved in the execution of the regicide, Damiens. *Source:* Courtesy of the Bibliothèque nationale de France, réserve QB-201 (106)-FOL.

that turned the scaffold into a stage, was ready-made for the principles of late eighteenth-century French sentimentalism.[52] An execution consistent with its moral imperatives would encourage dignity and patriotism; one that reproduced the horrors of the Ancien Régime would dull empathy and dissolve the common bonds that linked citizens and humanity.

The Unfailing Machine

The guillotine was designed to eliminate the threat of public horror that failed executions posed for the sentimental self. The device emerged from a collection of proposed reforms to capital punishment that, initially, had nothing to do with machines.[53] When Joseph-Ignace Guillotin first spoke on the question of capital crimes before the National Constituent Assembly on October 9, 1789, it was during a debate on penal code reform. The exact text of Guillotin's speech is lost to history, if it was ever recorded. But if the parliamentary records are accurate, he initially made no mention of a decapitation device. Guillotin instead proposed to add six articles to the penal code—a set of interlocking legal reforms that embedded criminal punishments within a tableau of contemporary judicial concerns surrounding the treatment of criminality writ large.[54] If the Assembly had simply adopted Guillotin's proposal that October, the French Revolution might never have seen a decapitation machine. But the legislators repeatedly demurred. Undaunted, Guillotin—a physician and freemason, a Jacobin, and a representative of the Third Estate—took up the question again in November. Discussion was again delayed. Finally, on December 1, his proposals came up for debate in the Assembly.

Contrary to what we have been taught, Guillotin was not primarily concerned with eliminating the physical suffering of the condemned criminal. His idol was the sentimentalist Diderot, who, under his entry for "anatomy" in the *Encyclopédie*, had proposed and defended the vivisection of criminals as humane. Diderot had reasoned that vivisection punished the inhumanity of the criminal and employed the full capacities of the surgeon for the benefit of all humanity.[55] His litany of proposed procedures (performed in private) included injecting different liquids into live criminals, allowing their thighs to be amputated at the joint, joining their mammary and epigastric arteries, sawing out their ribs, or having a portion of their brain extracted. Here, criminal suffering was not only an afterthought; it was squarely compatible with sentimentalist principles. In 1775–1776, in a memorandum on the treatment of rabies, Guillotin took up Diderot's idea, proposing that convicts should undergo "all such experiments as have been . . . attempted with animals" as a way of reintegrating them into society. The procedure may seem "unjust, cruel, terrifying or unnatural," Guillotin explained, "but it is merely alarming."[56] Like so many of his contemporaries, Guillotin was concerned with the profitability of executions within a larger moral economy of punishment. Shortening the suffering of the condemned was a secondary benefit.[57]

Instead, Guillotin's interventions aimed to eliminate the effects of pathological emotion on the sentimental selves who *witnessed* executions—friends, family members, the crowd.[58] Enlightenment arguments for humane punishment routinely expressed the effects and the horrors of Ancien Régime punishments in the first person and as a discourse rooted in the self.[59] Pierre Louis de Lacretelle, a royalist historian and journalist from Metz who beat out Robespierre in the Académie's 1783 essay contest and who had drawn up the *cahiers de doléances* with Guillotin, commented: "Thank God for those gentle, sensitive souls on whom those horrible executions exert a kind of torture."[60] Guillotin began his own December proposal by detailing the farrago of horrific criminal punishments that characterized Ancien Régime justice: highwaymen were broken publicly on the wheel, thieves were hung, regicides were quartered, counterfeiters were boiled alive. Only beheadings were reserved for the nobility; and even these could be shocking, as the cases of Damiens and Lally had shown. The *Journal des États généraux* provided a chorus for his views, describing how "Monsieur Guillotin dwelt at length on the tortures in which man shows himself more ferocious than wild beasts. The torment of red hot pincers and such like, these things I pass over in silence."[61]

Guillotin's original proposal in October had sought to rationalize these practices, culling the variety of punishments and producing a single penalty for all capital crimes. But his new proposal in December also drew heavily on sentimentalist concerns over public horror and the place of machinery in eliminating them. Maximizing the dramatic effect, Guillotin reorganized the sequence of his original recommendations, saving the most striking suggestion for last. The first five articles now protected the criminal's property, family, and individual rights, as well as establishing equality of punishment across rank and station. Only the sixth article treated capital punishment specifically, doing away with the different regimes of execution and specifying that all death sentences would share a single method—decapitation—carried out by a "simple mechanism." Against the chilling, chaotic tableaux of Ancien Régime executions, Guillotin sketched for his fellow revolutionaries a new vision: a uniform method of punishment for all capital crimes, and a dependable simple machine to carry it out. "The mechanism falls like thunder, the head flies, the blood flows, the man is no more."[62]

Guillotin's mechanical vision drew explicitly on sentimentalism's politics of breakdown. He rejected the vice of complexity and its tendencies toward malfunction while embedding the Revolution's virtues (and his own) in the simplicity of the new device. His proposal extended that politics into new and perilous territory, conceiving mechanical failures not just as signs of immorality but as threats to social order. In doing so, he argued that sentimental selves would feel the breakdowns of the machine.

Because he understood this capacity to reach into the interior lives of spectators, Guillotin envisioned a sentimental cordon around his decapitation machine—a precinct that stopped the spread of pathological emotion. The cordon protected spectators not only from the emotional effects of execution but also from the possibilities of malfunction. His 1784 report on mesmerism, coauthored with Benjamin Franklin and Lavoisier, among others, had worried about the diffusion of potentially dangerous sentiments, particularly in the emerging public spaces of late eighteenth-century France: "Little by little the impressions spread and reinforce one another. . . . This sign of individual emotions establishes a general emotion that each person shares to the degree that they are susceptible . . . men gathered in large numbers are more susceptible to their sentiments [*sens*], reason has less sway over them."[63] That fear ran as an undercurrent through his proposal on capital punishment, where he suggested abolishing public executions altogether. And it haunted the possible breakdowns of his proposed device. An early illustration, reproduced from his description before the National Assembly, captured the carefully constructed precinct of sentiment, centered on the machine (figure 1.3). Its caption explained how executions would be performed outside the city in a specially designated location, ostensibly to limit the crowds. They would be rare and private, even intimate, affairs. To keep spectators at a distance, barriers would surround the machine and soldiers with lowered weapons would guard it. The confessor would give the signal for death at the instant of absolution. Even

Figure 1.3
"Machine proposed to the National Assembly for the execution of criminals by Mr. Guillotin" (1791). Note how the executioner turns away at the moment the rope is cut. Note also the use of the sword, the traditional method of decapitation. Courtesy of the Bibliothèque nationale de France, département Estampes et photographie, réserve QB-370 (29)-FT 4 [De Vinck, 4987].

the executioner, yielding to natural sentiment, would look away at the final moment.[64]

The National Constituent Assembly ultimately acknowledged Guillotin's anxieties, although not in the way he had hoped. It approved his proposal for decapitation as the punishment for all capital crimes. But over the course of the next month, into January 1790, the legislative body essentially uninvented the guillotine by undoing two central elements of its original vision. They rejected the idea of a decapitation *machine* (therefore calling into question again the exact method of decapitation), and they insisted on the social value to be extracted from *public* executions.[65]

Separating the dangers of mechanical action from public spectacle, the Assembly officially adopted decapitation as the universal form of capital punishment on June 3, 1791 (Article 3), insisting on its educational value. As late as March 1792, though, the method of decapitation was still assumed to be the sword. Asked for his opinion, the hereditary Royal Executioner of France and later High Executioner of the Republic, Charles-Henri Sanson, stressed that swords would quickly dull under multiple executions, requiring a store of spares. Anticipating the steady flow of executions in the years to come, with executees overcome by fear and unable to compose themselves, he cautioned that the event could become a struggle and a massacre, with possibly disastrous effects for executioner, for the victim, and, above all, for the public.[66] Even the executioner might be moved (as in Guillotin's depiction), and in such a delicate operation, emotions were dangerous. The house of Sanson was filled with accounts of pity, clemency, reluctance, and discretion—of the executioner even weeping while performing his duties. It was also tainted by a recent history of incompetence, as Sanson had proved himself inept at both hanging and decapitation.[67]

The resulting legal limbo—uniform capital punishment without an official method—created a powder-keg situation. Anyone convicted of a capital crime was sentenced to death; but executions couldn't be carried out until the law specified a method. By early March, the situation had reached a fever pitch. Murders were multiplying. The *commissaires du roi* (regional judges) outside Paris feared that public order was dissolving and that convicted criminals would turn increasingly violent while they awaited their fate.[68] By the first week of March 1792, the directory of the *département* of Paris could no longer contain its anxieties about both the need for an execution method and the consequences of its potential failure. It feared that "by the imperfection of the means or a lack of experience and clumsiness, the ordeal would become horrible for the patient [*sic*] and for the spectators." The *procureur-général*, Pierre Louis Roederer, echoed those concerns. "The executioner, from a lack of experience, could turn the decapitation into a dreadful [*affreux*] torture, which is what we are trying to stop." Sanson himself agreed. For the execution to follow the intention of the law, the executioner had to be very skillful, and the condemned very firm. Otherwise, one could never finish an execution by sword without producing "dangerous scenes."[69]

With pressure mounting, the Legislative Assembly turned to France's foremost expert on death to help decide the question. Antoine Louis, Perpetual Secretary of the Academy of Surgery, had served for a time as a military surgeon in the Army of the Lower Rhine. His extensive experience with amputations alone might have qualified him as an authority. But two additional elements ensured his selection.

First, in 1752, Louis had become embroiled in a debate over the certainty of signs of death, sparked by a panic over recent, sensational cases involving revived drowning victims who were presumed dead and buried alive. His treatise, *Lettres sur la certitude des signes de la mort*, aimed to reassure the public against fears they could suffer the same fate.[70] This expertise led Diderot to ask Louis if he would author the *Encyclopédie*'s entry on "death." The second factor, was Louis's place in surgery's recent rise from artisanal trade to liberal art and its relation to both mechanical devices and the state. As its French name implied, surgery was a form of manual labor in antiquity—the term *chirurgie* derived from the Ancient Greek χειρουργός (*kheirourgós*), meaning "the work of the hand," and denoted the branch of medicine that cured by manual interventions.[71] Lamenting their association with low-status barbers, many French surgeons in the early eighteenth century began casting themselves as *artistes*—a group that included painters, sculptors, pharmacists, clockmakers, and metallurgists, and who combined the learnedness and politeness of the savant with the usefulness of the artisan for the state.[72] Following a series of running battles with the powerful and condescending Faculty of Medicine, the "surgeons of the long robe" succeeded in founding their own royal academy in 1731 and in having surgery recognized as a liberal art two years later.[73] As Perpetual Secretary of the newly powerful Academy of Surgery, Louis's place in the history of the profession—its identification with the *artiste*, its battle with the physicians, its furnishing of state experts, and its connections to the mechanical arts—helped cast him as an authority on both death and the machines that would most effectively produce it.

As secretary, one of Louis's most important responsibilities was to certify the sentimentalist-inspired machines submitted to the Academy for approval. One criterion he repeatedly invoked in his evaluations was whether patients would *sense* the operation of the machine. In late 1778, for example, he registered the approval of a mechanical bed, demonstrated by the *architecte méchanicien* Mathieu, which allowed patients to be turned and lifted "without causing them any *sensible* movement." He would emphasize the simplicity of the machine, "as simple as it is ingenious," as a virtue and a prediction of its proper functioning.[74] That theme of simple machines, their protections against failure, their tendency to go unfelt when they worked, and their ties to sentimentalist virtues and public order would inform Louis's proposal for a method of decapitation.

Louis submitted his official report to the assembly on March 7, 1792. It began by surveying historical methods of decapitation while underlining their effect on the sentimental spectator. There is reason to believe that Louis had witnessed firsthand the botched execution of Lally, and he now used it to illustrate the difficulties of traditional methods. "[Lally] was kneeling, his eyes blindfolded; the executioner struck him at the nape of the neck; the blow failed to separate his head and could not have." It took three or four additional blows, administered by Charles-Henri Sanson's father, who took over the execution, to finish the act. Invoking the sentimental self, Louis cringed: "One watched this butchery with horror." Describing methods used in Germany and Denmark, Louis explained the mechanical and physiological difficulties involved: dense bone, cartilage, the fallibility of human executioners. As a surgeon, he knew firsthand the problems that varying human skill and, in an age before anesthetic, the struggles of the executee posed—echoes of Sanson's fears of massacre on the scaffold. Louis soon concluded that it was impossible to achieve a quick and certain decapitation "by confiding it to an agent who is susceptible to variations in ability, due to moral or physical causes." He continued, "It is absolutely necessary, for the certitude of the procedure, that it depend on invariable mechanical means whose force and effect can be determined." More than two years after Guillotin had first proposed it, then, Louis returned to the idea of a decapitation machine. He made clear that it was in light of the terrible spectacle of previous methods that "the present machine had been imagined."[75]

Louis went further, though. In a statement rife with ambiguity, he suggested that the success of the machine lay once more in its ability to go *unfelt.* "This device, if it seems necessary, will cause no sensation, and will barely be noticed." In perhaps the most remarkable statement of an already remarkable report, Louis went on to assert that it was "an easy enough matter to build such an unfailing machine."[76]

But what did infallibility look like in the technological imagination of the early French Revolution? In retrospect, the form of the guillotine strikes us as obvious. But as Ken Alder has explained, technical requirements underdetermine any machine's final form.[77] Others could, and did, imagine different decapitation machines, not all of them informed by sentimentalist principles or its concerns over breakdown. The Alsatian notary, Laquiante, had proposed a simple but flimsy device, like a double-hung window in which two blades met like scissors, almost ensuring breakdowns and horrific spectacles. A German depiction (figure 1.4), published in Augsburg after 1792 and produced from rumors and secondhand accounts of the French device, also imagined a very different, more complicated machine: a kind of cuckoo clock laying on its side. A mechanism inside the stage raised the criminal through an opening in the floor, where a spring-driven saber dispatched them. An almost identical German depiction claimed (falsely) that this was, in fact, the guillotine in operation in Paris, capable of executing fifteen victims per minute or more than ten thousand per day.[78]

Guillotin himself had proposed a noble, neoclassical machine in line with his views about the dignity of the condemned. Louis, with his links to the world of the mechanical arts, would instead commission an artisan's machine. His account traced age-old historical methods of decapitation going back to Ancient Rome, a long genealogy of execution machines that passed through the Italian *manaia*, the Halifax Gibbet, and the Scottish Maiden. But at the moment that Louis sought to specify the form of the French machine, his genealogy suddenly cut orthogonally to the urban geography of contemporary Paris. He turned to what many viewed as a perfected mechanism—one that filled the construction sites of the French capital: "From the height of a cross-beam that joins the two posts one drops, by the means of a release [*déclique*] . . . the convex axe whose back should be strong and heavy enough to act effectively, like the *mouton* used to embed piles."[79]

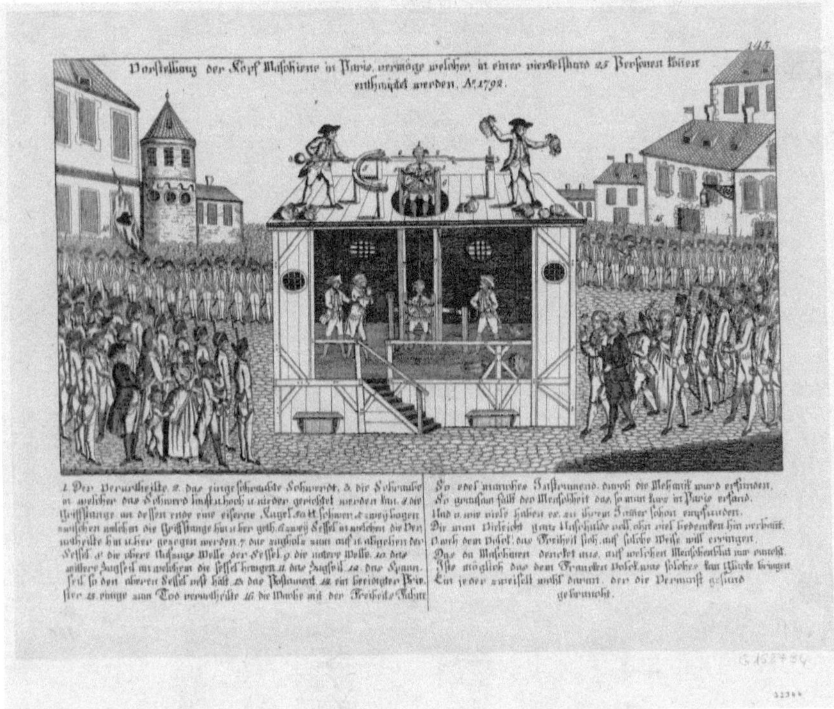

Figure 1.4
German representation of the guillotine. The title reads "Presentation of the decapitation machine in Paris, by means of which, in one quarter of an hour, twenty-five people can be decapitated." Vorstellung der Köpf Maschiene in Paris, Bibliothèque nationale de France, département Estampes et photographie, RESERVE FOL-QB-201 (129). *Source:* Bibliothèque nationale de France.

The *mouton* was already an old mechanism in the late eighteenth century and a centerpiece in Paris's public theater of machines. A falling weight guided by two uprights, it was a mainstay of early modern manufacturing and the building trades. Mid-century treatises, including the *Encyclopédie*, placed the mechanism within a pantheon of construction devices that included cranes and scaffolds (figure 1.5). It featured in the pile drivers used to set the *pilotis* for the bridges spanning the Seine. On January 20, 1791, it drove the first piles for the foundations for the Pont Louis XVI, which crossed the river between the Bourbon Palace and the Place Louis XV (now Place de la Concorde).[80]

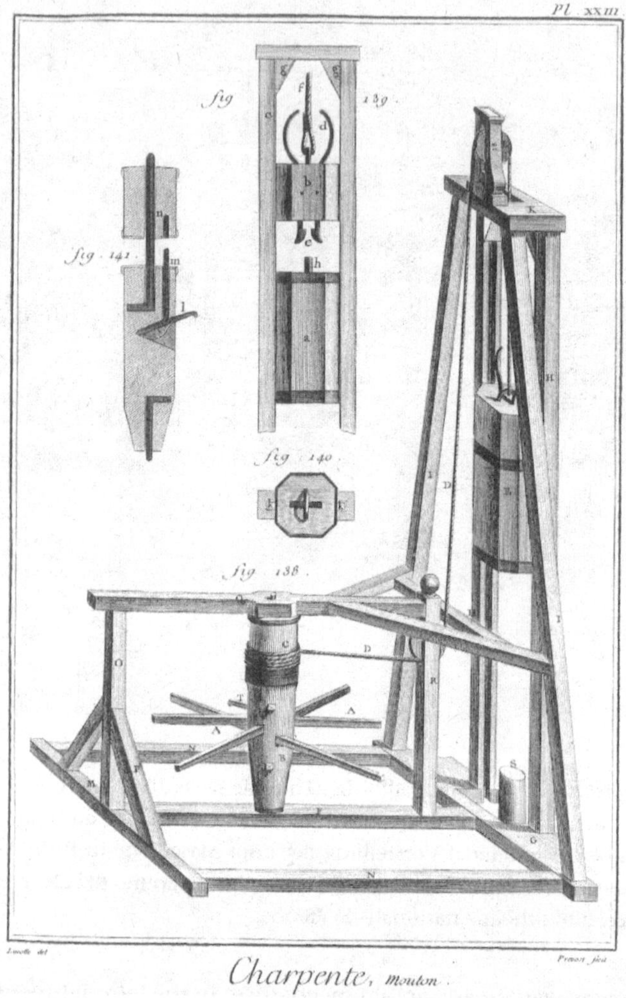

Figure 1.5

The *mouton*. From the *Encyclopédie* entry on carpentry. The falling weight labeled "a" in figure 139 (center top), and "E" in figure 138, represents the *mouton* proper, but the term was often used to refer to the entire mechanism—uprights, pulleys, cord, and falling weight. *Source:* «Charpenterie», *Encyclopédie ou Dictionnaire raisonné des sciences, des arts et des métiers*, vol. 2 (pl. XXIII), Paris, 1763.

During the two-decade building boom that preceded the Revolution, and which rebuilt fully one-third of the capital, the *mouton* was seemingly everywhere.[81] It formed part of the urban soundscape. Louis would have likely heard its percussions from his study as he wrote his report. It helped erect public buildings and private residences. At least one of the devices, set up in the fall of 1789 at the future national workshops of Jean-Joseph Dauffe, carried the nation's seal stamped on one of its uprights.[82] Dauffe originally used it to create jewelry and buttons for soldiers' uniforms, but as the Revolutionary Wars intensified, his workshop began using it to stamp gun locks for the rifles that the French citizen armies would carry into battle.[83] On the other end of the city, the *mouton* figured in the crises of revolutionary finance. On Rue Sainte-Anne, the Commission des Monnaies used it to stamp the plates for the *assignats*, the revolutionary paper currency introduced by the National Assembly and secured against Church lands to avoid bankruptcy in 1790.[84] Used to pay the large workforces of the construction trades, the *assignats* underwrote the large-scale construction projects that deployed the *mouton* so prolifically, bringing its connections full circle.[85] By March 1792, when Louis imagined an infallible machine, the *mouton* formed, quite literally, a machinery of state: building the public and private infrastructure of the capital, ensuring the solvency of the revolutionary nation, manufacturing the weapons that French armies would carry into battle, and, in Louis's vision, asserting the ultimate authority of the French state over its citizens in the form of the new decapitation machine.

Building the Unfailing Machine

On March 20, 1792, the Legislative Assembly, anxious about the prospects of failed executions and sensing insurrections materializing around them, adopted the mechanism outlined in Louis's report as the official method of decapitation in capital crimes. It authorized the funds for construction and mandated that it should be uniform throughout all of France.[86]

But even before the Assembly had officially approved the decapitation machine, the *procureur-général*, Roederer, had summoned Guillotin in early March to his elegant offices of the *directoire* at Place Vendôme number 4 (possibly to ask for insights on the construction of the device, possibly to ask Guillotin himself to oversee the project).[87] Guillotin apparently refused any further official involvement. Roederer then began working urgently with Antoine Louis in the third week of March to recruit a dependable and conscientious artisan. With capital cases piling up and a lucrative contract to supply all of France on the line, artisans began freely offering their ideas and services. On March 24, Louis wrote to Roederer about one prospect in particular. An émigré German piano maker had come to his attention, who believed he could "exercise his genius" in constructing the device.[88]

Just how Tobias Schmidt came to Louis's attention is something of a mystery.[89] Despite Schmidt's activities in the heart of the Cordeliers, where Louis himself lived, it appears he and Louis had never met. The Paris marriage registry confirms that Schmidt was born on January 30, 1755, in Kloppenheim, Germany (near Weisbaden).[90] We know that by 1785 he was already in Paris, since he was listed in the "réception de maîtres et maîtresses de métiers" for July 28 of that year.[91] In 1788, on the eve of the Revolution, he was living at Saint-André-des-Arts, number 54, around the corner from his workshop on Cour de Commerce.[92]

Schmidt was variously described in contemporary sources as *artiste*, *méchanicien*, and *artisan*. He would describe himself as a piano maker who occasionally abandoned the art to invent "mechanical discoveries useful to humanity."[93] Early on in the Revolution, he would propose the diving machine that began this chapter, a plow that could be pulled with half the normal force, and a ladder that might rescue fire victims unable to descend stairs. He shared the widespread commitment to mechanical simplicity that characterized the period, advertising his new pianoforte by explaining "the construction of this instrument is very simple," and that apart from the bascule, there was no friction.[94]

However it happened that Louis met Schmidt, the artisan proposed one possible design for an execution machine in the days following the Assembly's decision. The concept seemed sound to Louis, and he endorsed the design but did not yet offer Schmidt the commission.

With Schmidt in reserve, the Directory government charged Louis with approaching the King's carpenter, Guidon (sometimes named as Guédon in the records). As *charpentier du domaine*, Guidon was charged with building the *bois de justice*, the execution scaffolds, in the capital and the surrounding area. The carpenter visited Louis on the morning of March 30, accompanied by one of his fellow artisans. The three men discussed the invention, came to what Louis called a "perfect understanding" of the device, and

Louis passed on detailed instructions to avoid any misunderstanding. His specifications stressed solidity and sureness:

1. uprights, composed of beams ten feet high, six inches thick, with reinforcing struts on the sides and behind and copper pulleys on each support
2. the blade "of good temper," and of the solidity of the best cleavers; its back will have the thickness of an axe, and will be mounted with a weight of thirty pounds, subject to modification by experiments
3. a rope of sufficient length and strength; a solid wooden billet
4. the stockade for holding the accused
5. the metal shackle for the neck

Inspired by the sentimentalist view of spectacle as pedagogy, however, Louis departed from Schmidt's design and made one crucial modification to the *mouton* that would arguably, and inadvertently, cause its first breakdown. The traditional pile driver that shaped Louis's vision of the guillotine had originally contained a pincer mechanism—a kind of metal tong, sometimes encased in a wooden block separate from the *mouton* itself (see again figure 1.5). When the wooden block was lowered onto the *mouton* proper, the tong would open and automatically latch onto a metal ring on the *mouton*'s upper surface. Block and *mouton* were then hoisted together, and once they reached the top of their travel, two slanted pieces of wood compressed the handles of the tongs, releasing the *mouton*, which fell, striking the wooden pile and driving it (over successive iterations) into the ground. The sentimentalist spectacle that Louis envisioned for the decapitation machine, however, required holding the blade at the full height of its travel, creating anticipation and suspense. Guillotin had achieved this by tying the rope directly to the *mouton*, bypassing the pincer and the wooden block (see again figure 1.2). The executioner hoisted the *mouton*, secured the rope once it reached its apex, and cut the rope with a saber at the appointed time. Louis kept the essentials of this arrangement but introduced a latch to hold the *mouton* in place and a second string or rope that the executioner pulled to activate the release. This double rope arrangement—one to raise the *mouton*, the other to trigger the latch—would feature in the debates over what, and who, caused the machine to break down.

Guided by Louis's design, Guidon presented his estimate the next day. The cost would include not only the device in all its details—ropes, copper pulleys and grooves, the *mouton*, three blades, scaffolding, and staircase—but also the trials of the device and a miniature model that would reveal any possible faults in the design. Bracing Louis for the cost, Guidon stressed the quality of the construction: new oak of the finest quality, cast copper to match. The total sum would run to 5,660 livres.[95] He reassured Louis that this first exemplar would help to calculate exact costs and prevent *événements*—literally "events" but figuratively understood as breakdowns—in cheaper future machines. He also explained to Louis privately that the stigma attached to this type of work made workers scarce and that horses and carriages were needed to raise the scaffolding, both of which inflated the price. As a final point and to assuage Louis's concerns over material failures, he explained that the machine would last fifty years.[96]

The finance minister and ally of Mirabeau, the Swiss-born Jacobin Étienne Clavière, rejected the estimate as extortionary. With the "unthinkable" Haitian Revolution once again gaining force after setbacks in November 1791, Clavière (the first president of the abolitionist Societé des Amis des Noirs) was in the midst of convincing the King to adopt the Jacobin decree of equality between free Black people, mulattoes, and whites in Saint-Domingue.[97] On April 9, he opened up bidding on the contract. The following evening, Roederer asked Schmidt to visit him in the Place Vendôme and granted him the commission. Schmidt agreed that he would claim no compensation if the proposed experiments on cadavers failed. To fairly pay him for the trials and prototype, Roederer agreed on a price of 960 livres.[98]

The press of time was enormous. Schmidt promised to deliver the machine on Saturday April 13. After initial tests on sheep at his workshop on Cour de Commerce, trials could be organized for Sunday April 14 at the Bicêtre hospital, a cluster of buildings shared with a prison, an asylum, and a hospice that had recently housed women suffering from venereal diseases and transferred from la Salpétrière. And on Monday April 15 or Tuesday April 16, the executions could take place.

Roederer, feeling the intense pressure and hearing nothing about the trials' results, wrote impatiently to Louis. The next day, April 18, he finally received word. Tested on three cadavers, the machine had "decapitated so cleanly that we were stunned by the force and the speed of its action. The functions of the executioner will be limited to pushing a rocker [*bascule*] that allows the fall of the *mouton* carrying the blade."[99] Schmidt would construct the blade for all eighty-three *départements* of France; it would, Louis assured Roederer, be easy to find carpenters for the rest. The next day, he wrote a final note saying that Guidon would have a fortified platform ready on Saturday, and the execution could take place two days later, on Monday April 23.[100]

Roederer predicted the execution would attract a crowd of curious onlookers. Incidents of machine breaking, where revolutionaries had smashed textile machinery and burned their shattered frames, had grown in the early years of the Revolution.[101] Nervous, Roederer asked the Marquis de Lafayette—hero of the American Revolution, commander of the National Guard, and himself only four months from prison—to order the gendarmes to protect the machine from the crowd. "The new mode of execution—punishment by beheading—is sure to draw a large crowd to the [Place de] la Grève, and you would be well advised to take measures so that no damage is done to the machine."[102] He suggested that Lafayette order his gendarmes to remain in place, surrounding the machine, so that it could be safely removed after the execution.

The date was eventually set for April 25 in Paris. The victim would be a highwayman, Nicolas Jacques Pelletier. On October 14, 1791, Pelletier and an unknown number of accomplices had robbed and killed a passerby on the Rue Bourbon-Villeneuve, stealing 800 livres of *assignats*. Pelletier had been awaiting his execution since January 1792. He was taken to the Place de Grève, dressed in a red overshirt, at about 3:30 p.m. As Roederer had worried, the novelty of the spectacle indeed drew much larger crowds than usual, amplifying the possible consequences of any breakdown. Although the machine functioned exactly as planned, its effect on the public was completely unexpected: "The people . . . were not at all satisfied: they saw nothing; the thing was too quick."[103] The journalist Duplain gave his impressions two days after Pelletier's execution. The people, his witnesses reported, called for a return to Ancien Régime executions, singing "bring back my wooden gallows."[104] The spectacle of suffering they called for would not have to wait long. The failures that Louis and his colleagues feared were weeks away, and they would draw a very different connection between breakdown and the self.

Failure's Children

In the weeks following Pelletier's execution, Roederer began suspecting that Schmidt's price for the machine had been inflated. He approached the departmental architect of the Seine, Pierre Giraud, to examine the decapitation machine being held under lock and key in a nearby warehouse. Giraud arrived on June 5, accompanied by the building inspector, Jonquet.[105] The two conveniently confirmed Roederer's suspicions: "This machine was made with such haste [*précipitation*] that one could obviously not give it all the sureness and convenience [*sûreté et commodité*] necessary in its movements." The authors signaled the public danger of its imminent failure, adding that the machine, "however well-conceived in itself, hasn't been carried to the degree of perfection that it's capable of, and that for the public's peace of mind, should be added."[106] They also identified specific defects: the *mouton* itself was merely nailed, rather than screwed, and might become detached from the blade; the grooves in which it slid could swell under certain conditions, slowing its fall; two additional *moutons* would be needed as immediate replacements in case of a breakdown; the grooves of the uprights should be lined with copper; and a footrest should be added to the plank. As Anne Carol notes, "All these recommendations attested to the persistent fear that the mechanism, reputed to be more reliable than a human, would fail and repeat the cruelty of Ancien Régime punishments."[107] Equally damning, in Roederer's eyes, was the estimated cost of the machine. Schmidt had already produced guillotines for Paris and Versailles and several other *départements*. Eighty-three *départements* remained. Subsequent "perfected" machines should only cost 500 livres, and Giraud had very conveniently identified another artisan who was willing to produce them at exactly this price.[108]

It was no accident that Roederer chose an architect and a building inspector to determine the faults in a machine built by a foreign artisan. The frantic construction boom that had rebuilt roughly a third of Paris since 1760 had furnished a model for the guillotine in the form of the pile driver. But it had also generated its own string of spectacular, controversial, and highly public building collapses. Architects, charged with inspecting building sites before, during, and after construction, blamed a system of corporate monopolies, "an army of dirty and wretched [often foreign] workers," and a lack of policing and regulation.[109] In doing so, they placed the integrity of scaffolds, cranes, and even the *mouton* itself between the traditional authority of master carpenters and immigrant workers, on one hand, and the *architectes-experts-bourgeois* appointed with ensuring public safety, on the other. The controversies cast breakage, malfunction, and collapse as a tool for distinguishing patriotic elite French expertise from immigrant avarice, insensitivity, and indifference.

Following Giraud's inspection, Roederer became increasingly convinced that Schmidt needed to be stripped of his contract. The question of breakdown formed the most important plank in his attacks. On June 7, he wrote to Clavière with Giraud's findings: "I felt the need to inform you . . . of these details, because it's possible that your arrangement with M. Schmidt might not prevent a savings like the one that presents itself." Whichever artisan ultimately held the commission, Roederer stressed, they should not be allowed to send the machines to the *départements* without first having them examined and certified by an expert and having the certification sent along with the machines.[110] Receiving no reply, Roederer wrote on June 28 to the new minister of finance, Beaulieu, whose own predecessor had lasted only five days in the post.[111] Beaulieu responded with great interest and ordered Schmidt to reply to the architect's report. As a precaution, Beaulieu asked Roederer to get an official proposal from the rival artisan, René-Noël Clairin, before canceling Schmidt's commission. On July 13, Roederer provided Clairin's estimate and stressed that Clairin would provide the machines painted and certified by an architect of the minister's choosing.[112]

In an impossible coincidence, the carpenter Clairin also lived on the tiny Cour du Commerce, a stone's throw from Schmidt's work shed.[113] Most mornings, Schmidt would have passed Clairin as he ducked through the covered entryway that led to the small street and his workshop at the other end. And most days, Clairin could have simply glanced up the street to see the German hard at work on his inventions. He would have heard the bleating of sheep when Schmidt's prototype was tested. Perhaps out of loyalty to the trade, perhaps through some unknown arrangement, Clairin promised that even if Schmidt was stripped of his commission, he himself would produce a portion of the machines that Schmidt had been contracted for. Schmidt would be paid for the machines, and for the remainder, Clairin would receive 500 livres apiece, which would include instructions for use and certification by an architect.[114]

Over the next two weeks, Roederer's position hardened and his view of Schmidt dimmed further. His mood must have reached a nadir when Schmidt appeared at the posh offices of the *directoire* on July 16, asking Roederer for a copy of Louis's original letter praising the successes of the Bicêtre trials. In Roederer's office, Schmidt made clear his intentions to use Louis's letter to seek a patent for the decapitation machine. In a letter to the minister, Roederer explained that Schmidt had not previously dreamed of taking out a patent on a device that he clearly did not invent, a device to which he "only made a few changes to the description of M. Louis." Roederer calculated that Schmidt was now simply trying to compensate himself for his losses. He doubted the National Assembly would agree to the patent but wanted to warn the minister to avoid any surprises.[115]

The machine's breakdowns now assumed their most prominent role. In the last execution of the week of July 16, the Paris guillotine failed to completely sever the neck of the condemned. Mobilizing the sentimentalist language that had inspired the guillotine in the first place, Roederer seized on the event, informing the minister of circumstances that "made felt [propre à faire sentir]" the imperfections of the machine. Roederer claimed that Schmidt had cut corners. The grooves in the machine's uprights had swollen from the moisture in the air, slowing the fall of the *mouton* and botching the execution. Furthermore, Schmidt had foreseen this *inconvénient* from the beginning and had proposed lining the grooves with copper. (That claim was incorrect at best and a deliberate lie at worst.) But avarice had trumped this concern over breakdown. Roederer explained that only Schmidt's "personal interest had led him to avoid this cost" and therefore condemned the machine to fail.[116]

Schmidt himself provided a rival explanation. In his lengthy response, he rejected Roederer's claim about the copper-lined grooves, which would not have prevented the wood from swelling anyway. He summarized: "If the last execution in Paris did not fulfill the intended goal, it had nothing to do with the imperfection of the machine, which wants for nothing; rather it was the lack of precaution on the part of the executioner, *who did not take care to join the two ends of the rope that suspends the mouton*, and to hold them in such a way that they didn't hinder its movement." This lack of attention, Schmidt claimed, caused the rope to catch between the blade and the *croissant* (used to secure the criminal's neck), preventing the *mouton* from failing with its full weight. "The public witnessed this fact, and the brother of the executioner, who visited me on 29 July, reported it to me himself." Requiring the executioner to conform to the instructions for the machine, Schmidt explained, would avoid similar *inconvénients*. "For the rest, it is up to the executioner, before the execution, to see if the machine is in good condition, to take all the precautions to ensure its every movement; if there are concerns, he should remedy them. In this way, even the smallest accident can never happen." He finished by aligning the failures of the machine with the qualities and character of its operator: "The machine will always meet its expected aim, as long as the required precautions are not neglected."[117]

In the months to come, there were more reports of the guillotine's fallibility, particularly under heavy use.[118] Schmidt was steadfast in his defense, but the memory of those events would ultimately haunt Schmidt in the years to come. On 6 Germinal, Year 4 of the Republic (March 26, 1796), the Minister of Justice wrote to the Minister of the Interior explaining that Schmidt should travel immediately to Brussels to build nine decapitation machines. There, incredibly, the artisan demanded to be put in an enclosed government building to protect him from "malicious insults."[119] It would be his last documented involvement with the device, but it wouldn't be the final trace that its breakdown left on his self-conception. Ten years later, in one of the final pieces of documentary evidence we have for him, Schmidt advertised his latest invention for the 1806 Paris Industrial Exposition—a "piano-harmonica" that returned to his roots in instrument making. He described the device as "a simple machine, well-built, and guaranteed in this case to last 10 years."[120]

Conclusion

The "unfailing machine," as Louis described it, was never just about sparing criminal pain. It was one of many mechanisms for ensuring that the sensitive selves of the Revolution felt properly, that they were disturbed only in the right ways, and that their moral sensibilities were protected from the shock of disorder. As far as the guillotine was concerned, that project failed. Its breakdown left the masses unmoved, even as it reproduced the horrors of the Old Order. The guillotine's breakdowns laid bare not only the early fragility the machine's operations, but also the fragility of the Revolution's emotional regime as Louis and others imagined it.

At the same time, the breakdown opened a space in which deeper questions of belonging and difference could be staged. The indifference of Parisian crowds marked them, in the eyes of Revolutionary elites, as dangerously insensitive, a rabble unable to embody the sentimental virtues required of true citizens. The guillotine's builder—Schmidt—was caught in the same process of differentiation. The breakdown helped cast Schmidt as a suspect figure: a foreigner, an artisan, and ultimately an outsider to the

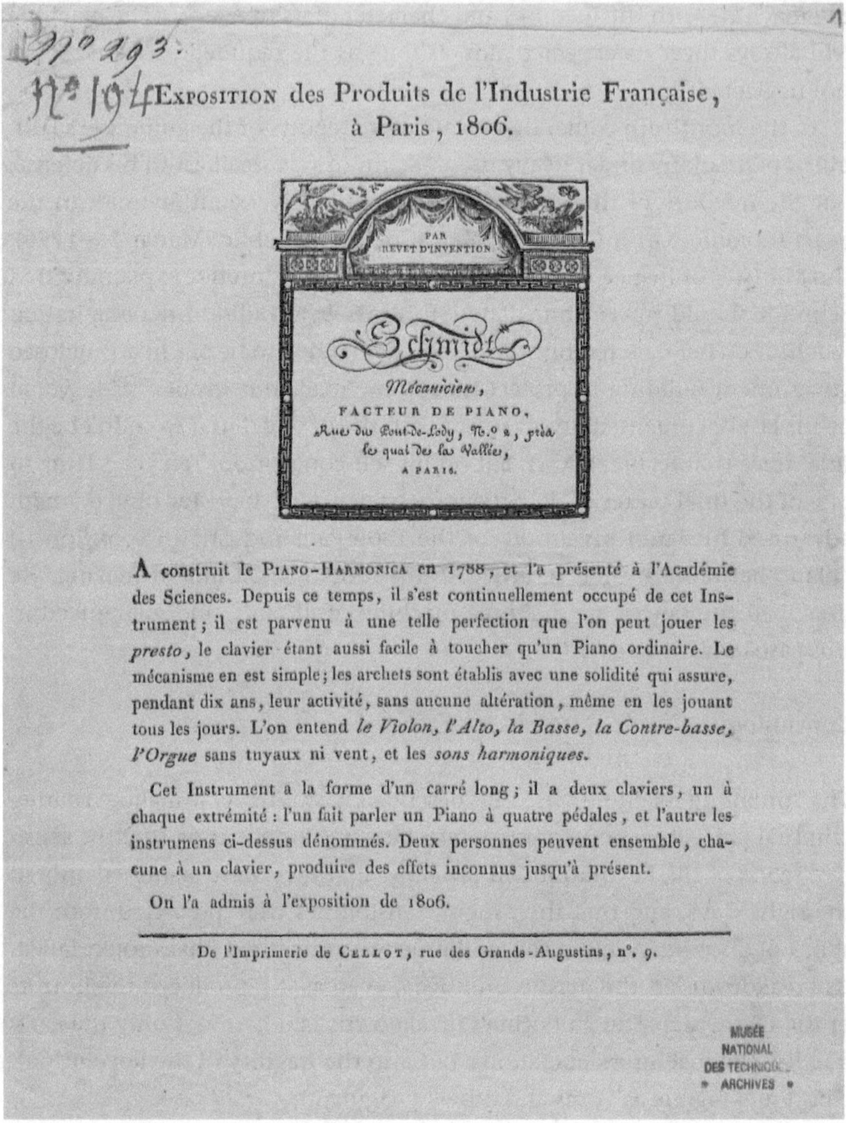

N° 293.

N° 194 EXPOSITION des Produits de l'Industrie Française, à Paris, 1806.

PAR
REVET D'INVENTION

Schmidt

Mécanicien,

FACTEUR DE PIANO,

Rue du Pont-de-Lody, N.° 2, près le quai de la Vallée,

A PARIS.

A construit le PIANO-HARMONICA en 1788, et l'a présenté à l'Académie des Sciences. Depuis ce temps, il s'est continuellement occupé de cet Instrument; il est parvenu à une telle perfection que l'on peut jouer les *presto*, le clavier étant aussi facile à toucher qu'un Piano ordinaire. Le mécanisme en est simple; les archets sont établis avec une solidité qui assure, pendant dix ans, leur activité, sans aucune altération, même en les jouant tous les jours. L'on entend *le Violon, l'Alto, la Basse, la Contre-basse, l'Orgue* sans tuyaux ni vent, et les *sons harmoniques*.

Cet Instrument a la forme d'un carré long; il a deux claviers, un à chaque extrémité : l'un fait parler un Piano à quatre pédales, et l'autre les instrumens ci-dessus dénommés. Deux personnes peuvent ensemble, chacune à un clavier, produire des effets inconnus jusqu'à présent.

On l'a admis à l'exposition de 1806.

De l'Imprimerie de CELLOT, rue des Grands-Augustins, n°. 9.

Figure 1.6
Advertisement for Schmidt's piano-harmonica at the Paris Industrial Exposition of 1806. Note the emphasis on reliability: "The mechanism is simple; the arches [*archets*] are made with a solidity that assures their activity for ten years, even playing them every day." Reproduced with permission of the Conservatoire Nationale des Arts et Métiers, P293.

project of the Revolution, even as he furnished its most iconic technology. This politics of breakdown became inseparable from the politics of identity. Just as the collapsed scaffolds and ruined buildings of the 1780s had been blamed on immigrant labor, and as French architects cast themselves as guarantors of order, the guillotine's malfunctions were read back into prejudices about who could feel properly and who stood outside the Revolution's sensitive community, dividing insiders from outsiders and recasting the very terms of Revolutionary citizenship.

Seen from this perspective, the guillotine emerges as something more complex than the "rational machine" that its mythology suggests. It was a sentimental machine, shaped by late eighteenth-century debates about the virtues of both mechanisms and the self, and about the proper management of feeling in public life. Its malfunctions mattered because they reverberated beyond the scaffold, unsettling the fragile equilibrium of emotions upon which Revolutionary order was imagined to depend. To trace the history of its breakdowns is to glimpse how precarious that order really was, how much labor went into cultivating expectations around the disturbance breakdowns should produce, and how spectacularly those efforts faltered. For Schmidt, the lesson was personal, leaving him with an enduring sense of how breakdowns shaped reputation and fate. For the Revolutionaries, it was political, teaching them that machines could never fully stabilize the volatile sentiments of a public in upheaval. In both cases, the guillotine's actions revealed how breakdowns were not only technical interruptions but moral and political events, shaping the politics of machines alongside the calculations of the soul.

2 The Danger of Systems: Victorian Railways and the Limits of Responsibility

Private Danger. Public Utility.
—Written on the side of the first passenger train in Britain, 1825

Late spring, the countryside in southeast England. On June 9, 1865, at exactly 3:13 p.m., Charles Dickens nearly died. And a concern over breakdowns was at its core.

Dickens was traveling with his mistress and her mother aboard a "tidal train"—so called because it would rendezvous with ships and kept their irregular schedule. Shortly after three o'clock, the train approached the first of three low bridges over the River Beult, near the town of Staplehurst. Unbeknownst to Dickens, his fellow travelers, or any of the crew, the bridge was missing a section of track. A work gang of carpenters and platelayers had been replacing the timbers that carried the rails over the viaducts east of the local station. These wooden elements formed part of the vast machine that Victorians called the "machinery of railways": the combination of rails, ties, switching points, sleepers, carriages, locomotives, and turntables that worked collectively to make rail transportation possible. Breakdowns in any one of the machinery's components could produce the spectacular railway accidents that swept through the popular press, threatened the traveling public, launched lawsuits and government inquests, and brought commercial and passenger traffic to a standstill. The workers were trying to avert this kind of materially inspired catastrophe by "renewing" the tracks.

Rather than interrupting traffic for repairs, the reigning practice on English railways in the 1860s was to carry out this maintenance in the intervals *between* passing trains. Timing this high-stakes work was the job of the foreman of platelayers. At breakfast that morning, the foreman had dutifully consulted his schedule book to calculate when repairs could begin. While scanning the timetable, though, he mistook the timing of the tidal train on June 9 for its timing on June 10. When a passenger train passed in mid-afternoon, he assumed it was the tidal train and that his men were free to carry out their work. (The crew's carpenter might have corrected him, but a few days earlier, a passing train had cut his schedule book in two.) Normally, warning detonators—small explosive charges, essentially large firecrackers—would have alerted the approaching train to the danger ahead. These were missing. By the time the locomotive driver saw the lookout, frantically waving a red flag just two hundred meters from the worksite, it was too late to halt the speeding train. The locomotive and two carriages successfully jumped the gap. But nine cars derailed, with one car lying astride the bridge and the dry riverbed below (figure 2.1).

Dickens was aboard that car. He was one of forty people injured and was nearly one of the ten dead. For two weeks afterward, he couldn't speak. Once his voice returned, he confessed that he still felt "a little shaken in my nervous system by the terrible and affecting incidents of the late railway

Figure 2.1
"Scene of the Fatal Accident at Staplehurst, on the South-Easter Railway—From a Sketch Taken the Next Day," 1865. *Source:* Illustrated London News, June 17, 1865.

accident." In the years that followed, the jolts of train voyages would send him into a panic, and he would grip the seats with both hands. His eldest daughter, Mary, recounted how "my father's nerves never really were the same again."[1] Five years to the day after the accident, Dickens died.

This chapter is not about Dickens or his shattered inner life. At least not directly. It is, instead, about the *kind of accident* Dickens experienced that day, its relationship to the material degradations that caused it and to a certain ideal of the Victorian self. Wolfgang Schivelbusch famously argued that railway disasters embodied the rise of the "technological accident": a new form of ominous calamity where technical objects destroyed themselves from within.[2] In this chapter, I argue that Victorian railways helped create an additional kind of accident that reconfigured the relationship between breakdowns, industrial modernity, and the self in Victorian Britain. That new class of accidents was largely the invention of a group of British Army officers drawn from the corps of Royal Engineers and tasked with investigating railway accidents throughout Britain. Into the 1840s, these inspectors generally relied on two tried-and-true methods for explaining why accidents happened. The first was to blame *things*—boiler explosions, broken axles, bridge failures, collapsed embankments, and so on. The second was to blame the inner lives of people, and particularly the *kinds of selves* they defined—the irresponsible, the inattentive, the lazy, the drunken, the stupid, and the reckless who populated the locomotives, switching points, and railyards of Victorian Britain, alongside their more "responsible" fellow Victorians. Starting in the early 1840s, however, an emerging class of accidents began defying these causal explanations and occupying a third category. By the time Staplehurst transformed Dickens into a nervous wreck, this new form of explanation was at its peak. These were accidents of the *system*, calamities that interwove machine failures—actual and imagined—with the interior lives of Victorians. And Staplehurst was just one of many.

The chapter explores how, between 1840 and 1880, in the hands of the railway inspectors, the "system accident" furnished a way to reconfigure the relationship between material breakdowns, personal responsibility, and blame in Victorian Britain. Its emergence helped answer at least two troubling questions of the age: who was responsible for the material degradations of industrial modernity and, by extension, who should be troubled (and how) by broken machines? I argue that the search to answer those questions was so pressing—riveting the attention of the public, spilling across the pages of newspapers, mobilizing the vast machinery of the Victorian state—precisely because railway breakdowns threatened more than Victorian lives and commerce, more than public health and safety, more even than the brute circulation of people and objects. They threatened long-established ideas about material and human causation and, with them, conceptions of self-responsibility, self-regulation, and self-governance that underpinned middle-class Victorian society. As such, they formed sites and instruments through which Victorians expressed what a person's experience of breakdowns said about their place in the changing industrial and social order of mid-nineteenth century Britain.

I also use that focus on breakdowns to revisit and revise our understanding of the history of technological systems. For historians of technology, the railway is often seen as the first modern technological system—a sprawling combination of heterogeneous material elements, increasing in scope and complexity, and interacting in potentially devastating ways.[3] Thomas Hughes's now classic formulation of "large technological systems" revised what we mean by that term, making clear how systems contain not only material artifacts but social and political components as well—all of which must be coordinated for systems to "work." [4] But that intervention raises the question of how technological systems like the railway came to be understood as primarily material formations in the first place, before Hughes's theory transformed them.

In the Victorian period, systems were already more than material assemblages. They were *schemes*: detailed methods for organizing collections of people and materials objects, shaping their movement and interactions, and ensuring social order.[5] I argue that the railway inspectors invoked this older and more abstract understanding of system and applied it to accidents in order to address a broad middle-class Victorian anxiety about the personal effects of industrial modernity. Located in the seat of British imperial power in Whitehall, the railway inspectors coined the system accident—calamities caused by flawed ways of organizing people and materials—partly to reconfigure the relationship between breakdowns and the Victorian self: shielding working-class railway employees from responsibility for breakdowns, absolving middle-class Victorians of blame for the consequence of industrial modernity, and stripping powerful railway companies of their previous immunity from accountability. The inspectors used systems to work out, for instance, whether a hapless and overworked foreman mending the tracks at Staplehurst should have to answer for the longer histories of extraction, degradation, and neglect that British industrial modernity had created. Under an intense popular backlash against the system as an explanation for accidents, these same inspectors materialized their abstract systems to help create the conception that Hughes (and others) could critique.

The chapter opens in 1844 by joining one of the railway inspectors as he travels to the remains of a terrible crash, one ultimately traced to a broken-down locomotive. Using that investigation, I explore the problem of responsibility that breakdowns posed for railway employees and inspectors, and how the rise of the "system" addressed it. The chapter then turns to the 1860s and the Staplehurst accident that began the chapter. As the volume of train travel exploded, medical authorities and the popular press began casting breakdowns as part of a wider threat to Victorian bodies and, tellingly, to their inner lives. Recasting the discourse over public threat that we saw in Revolutionary France, Victorians presented railway breakdowns as an inescapable disturbance in the modern self. Ironically, the new explanatory role of systems magnified and multiplied this disastrous effect of breakdowns, transforming them from occasional physical catastrophes to constant psychic disruptions. The closing section turns to a final accident at Princes Risborough in 1878 to show how, under widespread criticism of the systems explanation, the inspectors ironically proposed to materialize systems, refocusing responsibility on workers, preferring the possibility of material failure over fallible humans, and helping to create the understanding of systems, and system accidents, that we have today.

To see those connections, though, we need the richness that only a historical episode can provide. We need another accident.

Breakdowns and the Responsible Self

Just after 5:00 a.m. on October 8, 1844—a Tuesday—two English locomotives rounded a blind curve near Boldon Lane, not far from where the River Tyne empties into the North Sea. Sunrise was still two hours away, and it was dark. The line linking Gateshead to South Shields was a double line, with two sets of tracks running in parallel. But the locomotives, for reasons we will come to, were traveling on the same track, head-on. Imagine each machine hurtling through darkness, hobbled by weak brakes, its billowing smoke nearly invisible in the predawn, a riot of noise and machinery.

One machine was a coal engine—the *Nelson*—a testament to the line's origins as a mineral train serving the mines at Tanfield. The other was a passenger engine—the *Leopard*—pulling a single car carrying twenty-nine butchers to the market at Newcastle. With about three hundred meters of darkness separating them, each engine driver suddenly glimpsed the other's weak, amber headlights in the distance. As they bore down on their opposites, the gap between them closing at about a hundred kilometers per hour, the drivers frantically sounded their whistles, reversed their engines, and, sensing they were about to crash, leapt from their respective locomotives along with their firemen. The resulting collision was relatively minor, ripping the driver's platform from the passenger train but leaving the coal engine intact. Both locomotives were still operational at this point and there were no major injuries.

In his panicked leap to safety, though, the driver of the *Leopard*, who had been pulling the load of butchers, neglected to turn off his steam, violating a standard precaution. With its engine still running, and with smoke and darkness still encircling it, the train began slowly reversing with the passengers still on board. The driver desperately gave chase. He managed to grab hold of a railing but could not pull himself aboard. After being dragged about fifty meters, he tumbled into the embankment. The train continued accelerating, reaching full speed—about sixty kilometers per hour—as it barreled toward its point of origin: the station at South Shields. The butchers, sensing disaster, began leaping from the speeding train. Nineteen jumped clear before it collided full-on with three parked railway cars. Eight remaining passengers were injured. Two of them, John Brown and Joseph Burrell, were trapped for some time in the wreckage and later died.[6]

According to a protocol established in 1840, the Board of Trade sent an official inspector, a Captain Joshua Coddington, to determine what had caused the accident. Coddington traveled to the site of the crash, carefully examined the wreckage, interviewed witnesses and employees, attended the local coroner's inquiry, and, six days later, submitted his official report. It had, he believed, all started with a breakdown.

Breakdowns had plagued the Brandling Junction Railway after years of neglect and deterioration. When, on the evening before the accident, a broken-down locomotive had put the regular passenger train out of service, the foreman had ordered the *Leopard* to take its place, setting in motion a fateful chain of events that culminated in the accident. According to Coddington's report, however, although the breakdown might have been the first link in the chain, it was not the *cause*. Complicating the chain of events, the foreman had given a second order for the driver of the *Nelson* to bring up another train scheduled for the same time. That order was relayed through three different people, who altered it slightly with each retelling. By the time it reached the *Nelson's* driver, it took the form of a command to fetch the carriage carrying the butchers, which was already returning on the same track, in the opposite direction, hitched to the *Leopard*. Coddington's official report, however, found that the workers and their garbled communications had also *not* caused the fatal events.

Instead, Coddington landed on a third option: *the system*. In order to understand what it meant in 1844 for a military-trained, middle-class railway inspector to blame a system, instead of a thing or a person, for a railway accident we have to step back slightly to see the wider social role that railway inspectors imagined for themselves.

The Board of Trade's railway inspectors were drawn exclusively from the corps of Royal Engineers of the British Army, making them key agents of the Victorian state and of the British Empire. As a corps, the Royal Engineers were responsible for planning and constructing fortifications, surveying terrain, and managing the large-scale infrastructure projects—roads, railways, ports, and later telegraph systems—of both imperial expansion and domestic liberal governance. Coddington himself, for instance, had worked on the Ordinance Survey in Ireland for several years before designing defensive works in Bermuda.[7] As a government inspector, he was part of a large and growing body of specialist government agents, operating just below the governing caste, responsible for examining the conditions of mines, factories, prisons, workhouses, and schools.[8] Like that larger collection of government officials, the railway inspectors' job was to enforce standards and regulations for public order and safety: inspecting lines and railway structures—bridges, embankments, cutaways, tunnels—before they were opened for traffic. The Royal Engineers were specifically chosen as the source for the inspectors because of their perceived independence from the powerful railway companies.[9] Starting in 1840, the Board of Trade drew a small number of inspectors from this larger group to investigate accidents. In 1844, Coddington was one of only two such inspectors—the second to be appointed after Frederic Smith—and throughout most of the nineteenth century, there were never more than three serving at a time. (The entire staff of the Board of Trade in 1840 was only thirty.)[10]

The inspectors cast themselves as moral agents responsible for a larger project: creating a safe Britain within a dangerous world.[11] As engineers, they cultivated a specific kind of self as the core of their identity. As Iwan Rhys Morus observes for Victorian engineering at large, "Being an engineer meant being a certain kind of character."[12] Samuel Smiles, who would go on to write a number of engineering biographies, captured that character in his wildly popular book, *Self-Help, with Illustrations of Character and Conduct,* published in 1859 but inspired by lectures delivered in 1845. For Smiles, the fate of the nation rested on the character of its individuals: "national progress is the sum of individual industry, energy and uprightness, as national decay is of individual idleness, selfishness and vice."[13] On this view, engineers were exemplars of Victorian self-hood: hard-working, self-made, and self-disciplined, embodying Evangelical Protestantism's emphasis on good works as a mark of divine grace.[14] Where railway engineers such as Isambard Brunel or Robert Stephenson were defined by their vision and daring, though, the railway inspectors saw their work as reigning in the more dangerous tendencies of that material ambition. As stewards of both the progress and perils of modern industry, their work was defined by the virtue of vigilance. And they projected that morality, in the case of accidents, onto the people and machines of the railway network more generally.

The railway formed a prominent site for this intertwining of machines, morality, and vigilance precisely because of its novel capacities for breakage and disaster. Nineteenth-century European observers saw in railways a new and frightening form of breakdown with dire implications for the scope of human responsibility. The same year of the Shields accident, 1844, the French *Encyclopédie des chemins de fer et des machines à vapeur* devoted a nine-page article to the subject of accidents. In a language that recast the Enlightenment ideal of machine perfection as a potential curse, it explained: "All man-made things are subject to accidents. By a kind of compensation . . . , the more these things are perfected, the greater the gravity of the accidents that happen to them." It then laid out the demands that these perfected machines placed on humans:

> That is why, without rigorous surveillance at all points, the most powerful and most perfected industrial means, that is, steam engines and railroad trains, can

cause the most grave and fatal mishaps. The mass of the objects they set in motion, the velocity they engender, their very power, once halted or turned from its proper objective, is transformed into a terrible agent of destruction. Steam power, while opening up new and hitherto unknown roads to man, also seems to continually put him in a position best compared to that of a man who is walking along the edge of a precipice and cannot afford a single false step.[15]

On British railways in the 1840s, this problem of day-to-day machine surveillance generally fell to locomotive drivers, whose daily labor cultivated its own relationship between breakdown, vigilance, and the responsible self. As one engine driver explained: "The recollection of the past has taught him [the driver] that everything is consequent, and nothing comes by chance. Neglecting to look after a little steel pin—a trivial thing, some may be inclined to think—and yonder magnificent iron steed, champing and foaming under steam, is imminently liable to an accident, that will reveal the character of the driver." He continued, underlining the sensitivity that drivers needed to show to even the faintest hint of malfunction: "Neglect small things, and doubt their power, and the engine will be stopped, even by a grain of sand in the packing, . . . by the flaw in a leaf of a spring, by a false stroke of the pump, by a slip of the wheel, by a loose cotter, for the want of one spot of oil, one glance at the gauge-glass, heedless of a single notice." When engines broke, it was not because materials had failed but rather because the driver "considered not the things that, trifling in themselves, were capable of damaging an engine, so that it had to be fetched off the road. . . . In all probability, half a turn of a nut, half a glance, would have saved that engine."[16]

This machine-focused morality served the railway companies well, since it often absolved them of culpability for defects in infrastructure and rolling stock. As the railway network expanded, the companies extended this logic of vigilance and labor responsibility across the broader workforce—to signalmen, station managers, guards, and shunters—blaming these "railway servants" for the breakdowns that they either caused through alleged ineptitude or that they failed to notice, as well as for the accidents that resulted.[17] Blaming workers, though, increasingly ran against a growing public hostility toward the railway companies themselves. The joint-stock companies that formed the railway conglomerates in the early 1830s and 1840s produced some of the largest industrial enterprises in the world.[18] The prevailing public opinion held that the companies were criminally indifferent to passenger safety and unwilling to take responsibility for accidents, happy to let their workers take the blame instead. Throughout the 1840s, the railway inspectors were one of the steadiest defenders of railway workers in cases of material breakdowns, arguing that companies were responsible for the larger material conditions of their technologies.[19] The system became the inspectors' primary instrument for reconfiguring the relationship between breakdowns and the responsible working-class self and holding the powerful corporations to account.

The Rise of Systems

Like so many of the transformations in Victorian society, this role for systems took shape in the torrent of paper that crossed the desks of the Railway Inspectorate in Whitehall. Under the Regulation of Railways Act of 1840, every accident involving injury in Britain produced two documents that directly involved the growing administrative apparatus of the Victorian state.[20] The first was an accident return. This was a brief written notice that the railway company was required to file with the Railway Inspectorate within forty-eight hours of an accident on their line. Somewhere around March 1841, these returns became standardized, requiring specific information for each accident (figure 2.2).[21]

The notices were compiled and published as part of the larger "statistical returns" on railways: hundreds of pages of uninterrupted numerical tables that read like the material flows of mid-nineteenth-century Britain—first-, second-, and third-class passengers, manure, minerals, cattle, sheep, pigs, ballast, parcels, luggage, grain, fish, lime, timber, stone, bricks, hops, linen, woolen rags, leather, oil in casks, molasses, soldiers, and so on (figure 2.3). They positioned accidents firmly within the larger canvas of Victorian circulation, freedom, and commerce. They were also a reminder that the railway first assimilated human travelers into a system for moving commercial goods.[22] As Coddington would show in the case of Shields, this was part of the problem.

The returns inscribed a Victorian preoccupation with personal responsibility. They grouped accidents not according to their nature (collisions, derailments, people being run down, etc.) but rather according to whether the people injured were responsible for what happened. That responsibility was reflected in their discussion of cause.[23] In most cases, the returns

18 REPORTS *to the* COMMITTEE *of* PRIVY COUNCIL,

Appendix. I. Returns of Accidents. 1844. London and Dover (South-Eastern).

LONDON AND DOVER (SOUTH EASTERN) RAILWAY.

SIR, London Terminus, April 3rd, 1844.

It is with extreme regret I have to report to you a fatal accident on this line, on Monday last, the 1st instant, to David Nunn, late under guard in this Company's service, whose skull was fractured from a severe blow on the back part of his head, supposed to have been caused from his having, whilst on the roof of the carriage for some purpose or other connected with the luggage, come in contact with one of the bridges. The accident occurred between the Maidstone Road and Tunbridge Stations; it is not known where precisely, as his fellow guard had not missed him until the arrival of the train at the latter station. The train was the 1 o'clock "up" from Dover.

I am, &c.
JOHN WHITEHEAD, Secretary.

S. Laing, Esq.
&c. &c.

North Midland.

NORTH MIDLAND RAILWAY.

RETURN of Accident occurring in the course of the Public Traffic.

Date.	Name of Person Injured.	Description.	Nature of Injury.	Nature of Accident.	Cause of Accident.	Remarks.
1844 April 1.	J. Bentley Wheelhouse, corn miller, of Shelf, near Bradford.	A passenger by a second class carriage.	Arm broken, some ribs broken, contusions on the head, back, and side.	Fall from the train whilst running between Royston and Barnsley Station, occasioned by the passenger leaping out of the carriage.	Lost his hat whilst looking out of the carriage. He then opened the door, got on to the step, and leaped from the train. The guard observed him leap out, stopped the train, and took him to the Station at Barnsley, where medical aid was procured.	On being taken up, he was found to be injured and bleeding in the head and back; he appeared sensible, and expressed great anxiety for the recovery of his hat.

North Midland Railway, Derby,
April 3rd, 1844.

PETER CLARKE, Secretary.

Figure 2.2
The accident return. Railway companies were responsible for filing these notices with the Board of Trade for any accident occurring on their lines that produced injury. Note the columns for the nature and cause of the accident. *Source: Parliamentary Papers* 39 (London: HM Stationery Office, 1846), 18.

pointed to one of two possible causes: a material fault (e.g., a broken axle or collapsed embankment) or a human error or misjudgment, often related to character (e.g., drunkenness, imprudence, arrogance, or poor judgment). Their explanations were generally short and terse, respecting the limitations of the standardized forms. Throughout the 1840s, though, in a growing number of specific instances, accident after accident featured tortured explanations that burst the tidy columns of the tabular forms (figure 2.4). They suggest that, in these particular moments, Victorians *did not know how* to talk about, or even think about, the relationship between human responsibilities and material faults. We will return to this.

The brevity of the accident returns contrasted sharply with a second class of documents that accidents produced: the railway inspector's official report (figure 2.5).[24] From the beginning of their work in the 1840s, the inspectors used their reports as commentaries on Victorian social order. The

284 STATISTICAL RETURNS *made to the* COMMITTEE

LONDON AND SOUTH-WESTERN RAILWAY—*continued.*

Goods of every description, enumerating and classing together those kinds for the Conveyance of which different Rates are charged.	Number of Trains for Conveyance.	Average Speed, including Stoppages.	Number and Quantity Conveyed. By the Company as Carriers.	By other Carriers.	Total.	Total Mileage of each.	Average Charge per Mile. (a).	Gross Receipt.
		Miles.					d.	£. s. d.
Cattle	1,047	12·79	967	..	967	58,773	2¼	492 3 10
Sheep			30,790	..	30,790	1,691,411	½	1,444 3 4
Pigs			242	..	242	6,691	⅛	14 2 3
1st Class, *viz.*—			Tons. cwts. qrs. lbs.	Tons. cwts. qrs. lbs.	Tons. cwts. qrs. lbs.		per Ton.	
Coal			3,074 10 3 0	..	3,074 10 3 0	44,642	2½*d.*	
2nd Class, *viz.*— Coke, stone, timber, bricks, hides, potatoes, long manure, crude metals, &c.			6,221 0 3 12	158 14 3 9	6,379 15 2 21	176,477	3	
3rd Class, *viz.*— Lime, cement, bone-dust, grain, eggs, fruit, cheese, &c. exports and imports			5,262 19 3 23	2,981 9 2 9	8,244 9 2 4	427,767	3½	
4th Class, *viz.*— Sugar in hogsheads, paint, beer in casks, tallow, pitch, cutlery, paper, &c., carriers' heavy goods			2,529 16 1 16	3,756 14 3 4	6,286 11 0 20	305,430	3½	26,451 7 11
5th Class, *viz.*— Hops, linen and woollen rags, leather, dry hides, wool, &c.			1,323 3 0 10	16 4 2 0	1,339 7 2 10	92,516½	4	
6th Class, *viz.*— Oil in casks, molasses, sulphur, saltpetre, earthenware in crates, wine, silk, meat, shell and other fish, glass, carriers' sundries			2,026 0 2 14	3,496 10 2 0	5,522 11 0 14	333,413½	4½	
7th Class, *viz.*— Single Packages			5,128 15 1 14	..	5,128 15 1 14	323,091½	5½	
								£28,401 17 4

(*a*) No separate charge.

ALFRED MORGAN, Secretary.

HALF-YEARLY RETURN from 1st July to 31st December, 1844.

Description of Passenger Trains.	Number of Trains of each Class.	Average Speed of each Class. Exclusive of Stoppages.	Including Stoppages.	Number of Passengers in each Class.	Total Mileage of each Class.	Total Number of Miles Travelled.	Average Fare of each per Mile. Toll.	Locomotive Power.	Total.	Gross Receipts.
Fast Trains.		Miles per Hour.							d.	£. s. d.
1st Class	1,256	28·50	26·54	43,794½	2,428,757½	68,452			3½	31,962 19 0
Servants				3,642½	250,541½				2½	2,394 6 6
Mail and Mixed Trains.										
1st Class	736	27·75	23·67	89,525½	3,050,519¾	34,224	No separate charge.		3	36,627 3 11
2nd Class				223,187	6,257,550				2$\frac{1}{16}$	53,165 7 6
3rd Class	4,760	24·75	21·42	27,336½	985,879	172,593			1¼	5,988 7 2
Soldiers (1st month)				764½	47,833¾				1¼	224 4 5
Ditto (last 5 months)				8,302½	560,757½				1	2,336 9 9¼
Slow Trains.										
3rd Class (1st 4 months)	668	18·48	14·66	47,189	2,187,456	31,062			1$\frac{1}{16}$	9,598 12 2
Ditto (last 2 months)				16,544	641,516½				1	2,687 2 8
					Excess Fares received in addition to the above					579 6 3½
				460,286					£	145,564 1 5

—	Number of Trains for Conveyance.	Average Speed, including Stoppages.	Number and Quantity Conveyed.	Total Mileage.	Charge per Mile. Toll.	Locomotive Power.	Total.	Gross Receipt.
							d.	£. s. d.
Horses	6,752	23·88	2,782	162,755½	No separate charge.		3½	2,352 10 0
Carriages			2,094	122,421			6¼	3,114 1 0
			Tons. cwts. qrs. lbs.				Average per cwt.	
Passengers' luggage			304 18 3 11	20,923			⅛	825 3 4
Ditto by slow trains	668	14·66	91 9 0 19	6,282¼			⅛	128 15 4
							£	6,420 9 8

Figure 2.3

Statistical returns of the Board of Trade. The accident returns were included in the larger statistical returns of the Board of Trade, highlighting the broader flows that surrounded passengers. Note, for instance, the good listed in the upper half of the table. *Source: Parliamentary Papers* 39 (London: HM Stationery Office, 1846), 284.

8 REPORTS *to the* COMMITTEE *of* PRIVY COUNCIL,

Appendix.
I.
Returns of Accidents.
1844.
London and Southampton.

SIR, Nine Elms Station, 1st March, 1844.

The unfortunate man who attempted to cross the line of railway at Wimbledon, on the 18th instant, respecting whom I have already written to you, died on the 26th; the coroner held his inquisition, and the verdict is "Accidental Death," with a recommendation that more care be taken.

The gates were shut, a lamp was alight, our gateman was in attendance, and called out to the deceased not to pass, but this he disregarded. The surgeon who attended him after the accident proved that he was in a state to betoken intoxication, to which alone can his extraordinary imprudence be attributed.

I am, &c.,
ALBINUS MARTIN.

S. Laing, Esq.
&c. &c.

Bolton and Preston.

BOLTON AND PRESTON RAILWAY.

RETURN of Accidents occurring in the course of the Public Traffic.

Date.	Name of Person Injured.	Description.	Nature of Injury.	Nature of Accident.	Cause of Accident.
1844 Feb. 26	. .	. .	No injury sustained beyond a few very slight bruises.	A passenger train ran against some luggage waggons standing upon the main line at the Bolton Station. The engine and a third class carriage which was next the tender, were somewhat injured, and thrown off the rails, but only to the extent of a few inches, some of the waggons were also damaged.	The approach to the station from the north is round a curve between retaining walls, and it is in consequence impossible for the enginemen to see more, than it may be, 200 yards a-head. The luggage train, consisting of 37 waggons, had arrived at the station about 20 minutes before the passenger train, but was delayed by a change of engine, and by shunting waggons, (from different parts of the train) which had to be left at Bolton, the remainder of the train being destined for Manchester. A porter employed at the station, fearing that the passenger train might arrive before the luggage waggons were got out of the way, ran along the line to meet the coming train, and he and two other men employed upon the line, gave a signal to the engineman to pull up, when he was between 300 and 400 yards from the place at which the luggage waggons were standing; there appears to be reason to suspect that the engineer had not diminished his speed so soon as he ought to have done, in approaching the station, according to the instructions issued to the enginemen, and the difficulty of pulling up after he received the signal to do so, was no doubt increased by a fall of sleet at the time. The man in charge of the engine which was attached to the luggage train, attempted hastily to start his train, on receiving an intimation that the passenger train was approaching, but owing to the sudden jerk the coupling chains broke, and several of the waggons were thus left in the way of the passenger train. It is intended immediately to put in an additional side line at the Bolton Station, to enable the luggage trains to draw out of the main lines, whereby the liability of such an accident recurring will be greatly diminished, if not wholly prevented. The points and crossings for this siding had indeed been previously prepared, and the execution of the work had only been delayed through the change of management consequent upon the recent amalgamation of the Bolton and Preston with the North Union Railway Company.

PETER SINCLAIR.

Figure 2.4

A return describing an accident that would be attributed to a system. Note the extended explanation under "cause," signaling the crisis of explanation caused by what inspectors would term "system" accidents. *Source: Parliamentary Papers* 39 (London: HM Stationery Office, 1846), 8.

forensic discussions of the reports mirrored aspects of a narrative form that had emerged in the London newspapers in the 1820s, where stories about multiple accidents appearing under a single headline gradually gave way to extended and detailed accounts of a single accident.[25] Focusing largely on the plight of victims and avoiding discussions of cause, the newspaper reports encouraged their largely middle-class readers to see themselves both as responsible moral agents confronting a declining aristocracy and as blameless for the larger perils of industrial modernity.[26] The inspectors'

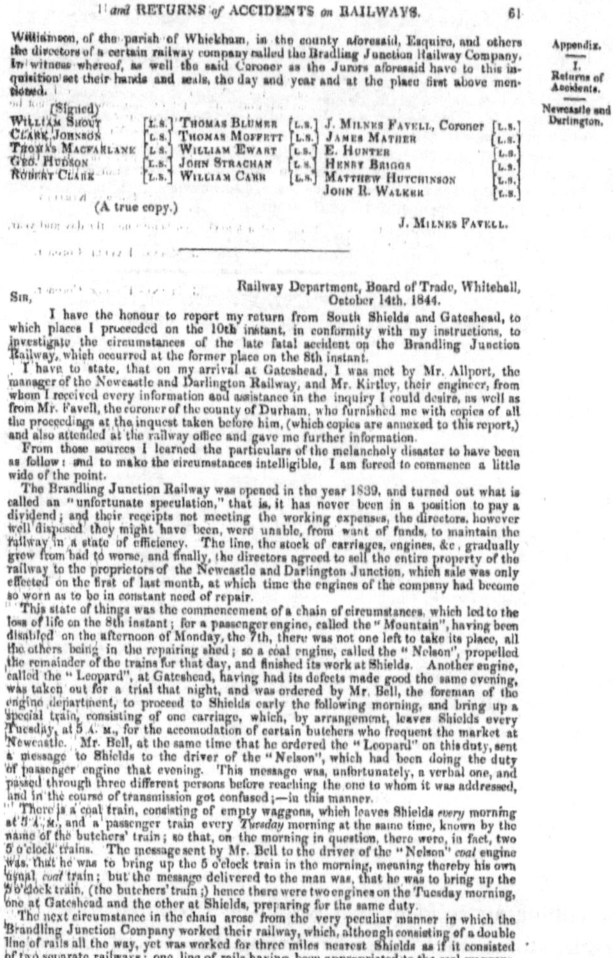

and RETURNS *of* ACCIDENTS *on* RAILWAYS. 61

Williamson, of the parish of Whickham, in the county aforesaid, Esquire, and others the directors of a certain railway company called the Brandling Junction Railway Company. In witness whereof, as well the said Coroner as the Jurors aforesaid have to this inquisition set their hands and seals, the day and year and at the place first above mentioned.

Appendix.

I.
Returns of Accidents.

Newcastle and Darlington.

(Signed)

William Shout	[L.S.]	Thomas Blumer	[L.S.]	J. Milnes Favell, Coroner	[L.S.]
Clark Johnson	[L.S.]	Thomas Moffett	[L.S.]	James Mather	[L.S.]
Thomas Macfarlane	[L.S.]	William Ewart	[L.S.]	E. Hunter	[L.S.]
Geo. Hudson	[L.S.]	John Strachan	[L.S.]	Henry Briggs	[L.S.]
Robert Clark	[L.S.]	William Carr	[L.S.]	Matthew Hutchinson	[L.S.]
				John R. Walker	[L.S.]

(A true copy.)

J. Milnes Favell.

Railway Department, Board of Trade, Whitehall,
October 14th, 1844.

Sir,

I have the honour to report my return from South Shields and Gateshead, to which places I proceeded on the 10th instant, in conformity with my instructions, to investigate the circumstances of the late fatal accident on the Brandling Junction Railway, which occurred at the former place on the 8th instant.

I have to state, that on my arrival at Gateshead, I was met by Mr. Allport, the manager of the Newcastle and Darlington Railway, and Mr. Kirtley, their engineer, from whom I received every information and assistance in the inquiry I could desire, as well as from Mr. Favell, the coroner of the county of Durham, who furnished me with copies of all the proceedings at the inquest taken before him, (which copies are annexed to this report,) and also attended at the railway office and gave me further information.

From those sources I learned the particulars of the melancholy disaster to have been as follow: and to make the circumstances intelligible, I am forced to commence a little wide of the point.

The Brandling Junction Railway was opened in the year 1839, and turned out what is called an "unfortunate speculation," that is, it has never been in a position to pay a dividend; and their receipts not meeting the working expenses, the directors, however well disposed they might have been, were unable, from want of funds, to maintain the railway in a state of efficiency. The line, the stock of carriages, engines, &c., gradually grew from bad to worse, and finally, the directors agreed to sell the entire property of the railway to the proprietors of the Newcastle and Darlington Junction, which sale was only effected on the first of last month, at which time the engines of the company had become so worn as to be in constant need of repair.

This state of things was the commencement of a chain of circumstances, which led to the loss of life on the 8th instant; for a passenger engine, called the "Mountain", having been disabled on the afternoon of Monday, the 7th, there was not one left to take its place, all the others being in the repairing shed; so a coal engine, called the "Nelson", propelled the remainder of the trains for that day, and finished its work at Shields. Another engine, called the "Leopard", at Gateshead, having had its defects made good the same evening, was taken out for a trial that night, and was ordered by Mr. Bell, the foreman of the engine department, to proceed to Shields early the following morning, and bring up a special train, consisting of one carriage, which, by arrangement, leaves Shields every Tuesday, at 5 A. M., for the accomodation of certain butchers who frequent the market at Newcastle. Mr. Bell, at the same time that he ordered the "Leopard" on this duty, sent a message to Shields to the driver of the "Nelson", which had been doing the duty of passenger engine that evening. This message was, unfortunately, a verbal one, and passed through three different persons before reaching the one to whom it was addressed, and in the course of transmission got confused;—in this manner.

There is a coal train, consisting of empty waggons, which leaves Shields *every* morning at 5 A. M., and a passenger train every *Tuesday* morning at the same time, known by the name of the butchers' train; so that, on the morning in question, there were, in fact, two 5 o'clock trains. The message sent by Mr. Bell to the driver of the "Nelson" *coal* engine was, that he was to bring up the 5 o'clock train in the morning, meaning thereby his own usual *coal* train; but the message delivered to the man was, that he was to bring up the 5 o'clock train, (the butchers' train;) hence there were two engines on the Tuesday morning, one at Gateshead and the other at Shields, preparing for the same duty.

The next circumstance in the chain arose from the very peculiar manner in which the Brandling Junction Company worked their railway, which, although consisting of a double line of rails all the way, yet was worked for three miles nearest Shields as if it consisted of two separate railways; one line of rails having been appropriated to the coal waggons, and the other line to the passenger traffic. The present proprietors appear fully aware of the dangerous tendency of this system, but as the line is unprovided with turn-tables, switches, and sidings, they were compelled to continue the same system as their predecessors until these necessary additions could be made, an operation which is in

Figure 2.5
Official accident report. *Source: Parliamentary Papers* 39 (London: HM Stationery Office, 1846), 61.

reports presented a similar vision of social order. Throughout the 1840s, in precisely the places where the shorter accident returns struggled to talk about cause, the longer official reports expressed that vision of social order by identifying "systems" as the culprit.[27]

We think we know what Victorians meant when they invoked systems in relation to railway accidents. That term conjures visions of complex and heterogeneous material elements and sprawling technological infrastructures.[28] In the English language, this understanding of systems as primarily material assemblages dates from about the second third of the eighteenth century.[29] It is generally *not* what Victorians meant by the term. Instead, Victorians repeatedly used an older understanding of systems with origins in the mid-seventeenth century: the system as an ordered scheme or plan of action; a regular method of procedure, government, or administration; "a scheme which unites many things in order," as Johnson's Dictionary put it.[30] (This is the root of the term "systematic.") Systems were a way of talking not primarily about collections of people or things but rather about the regularities and prescriptions for how these elements should work together, how much latitude they were given, and what roles they should occupy.

In drawing on this language in the 1840s, railway inspectors were participating in the rise of systems as instruments of Victorian governance, rooted in a local transformation that saw Whitehall become the center of the vast machinery of the Victorian imperial state (figure 2.6).[31] An explosion of administrative "systems" helped drive that concentration of government power: public health systems, penal systems, tax systems, census collection systems, systems of weights and measures. These were characterized by their scale, ambition, and complexity.[32] Systems paired abstract schemes with material devices—accounting books, statistical forms, metrological tools—that reigned in the behavior of humans, including government agents, ensuring their dependability while delimiting their responsibilities.[33] Over the course of the nineteenth century, those devices joined a number of other technologies—wide streets, sanitation infrastructures, telegraphy, public lighting—as methods of both cultivating and governing Victorian liberal subjects: the free, self-monitoring, and self-governing individuals who allowed for minimal state involvement in their lives.[34] As Tim Crook has noted, rather than acting as a monolith, the Victorian state itself operated through these diffuse arrangements and made them central to the problems of governing Britain and its empire.[35] They were one

PUNCH, OR THE LONDON CHARIVARI.—SEPTEMBER 26, 1874.

RAILWAY RESPONSIBILITY.

MR. PUNCH. "NO, NO, MR. DIRECTOR, *THEY*'RE NOT SO MUCH TO BLAME. IT'S *YOUR* PRECIOUS FALSE ECONOMY, UNPUNCTUALITY, AND GENERAL WANT OF SYSTEM THAT DOES ALL THE MISCHIEF."

Figure 2.6

"Railway Responsibility," 1874. The cartoon depicts the famous figure of Mr. Punch (left) and captures how "systems" reconfigured responsibility for accidents, shifting blame from workers to railway companies. *Artist:* Joseph Swain. *Credit:* The Print Collector / Alamy Stock Photo.

venue through which Victorians reimagined and restructured their society, economy, health, and political order, underpinning visions of progress and even utopia.[36]

Coddington's report used systems in precisely this way. To set the stage for his findings, he explained, he felt forced to "commence a little wide of the point." He began his report with a brief history of the Brandling Junction Railway. It had originally opened in 1839 to take over a wagonway dating from the 1720s, which brought coal from Tanfield to Staiths—staging platforms for loading coal—on the River Tyne at Dunston. Despite a surprisingly high demand from passenger traffic, the railway never turned a profit. The line and rolling stock gradually grew "from bad to worse," until the owners finally decided to sell the entire company in early September 1844. By then, "the engines of the company had become so worn as to be in constant need of repair." Coddington explained that "this state of things was the commencement of a chain of circumstances, which led to the loss of life on the 8th instant."[37]

Rather than blame the initial breakdown or the engine driver, though, which would have been customary, Coddington's narrative used the events surrounding the accident to reconfigure notions of responsibility and to shield the Brandling Junction's workers from the effects the breakdown would otherwise have on their responsibility and identity. He located the critical cause of the accident in "the very peculiar manner in which Brandling Junction Company worked their railway"—its *system* of operating. Most railways with double lines, such as the Brandling Junction, dedicated a single line to all traffic moving in the same direction: all eastbound traffic, whether passenger or freight, on one line; all westbound traffic on the other, for instance. The Brandling, however, probably owing to its early history as a mineral railway, worked the railway for three miles near Shields with all passenger traffic on one line (in both directions) and all coal traffic on the other (in both directions). As Coddington noted, the company's owners and officials appeared "fully aware of the dangerous tendencies of this system," but because the line lacked turntables, switches, and sidings, the arrangement persisted. These "defective arrangements" combined with "defective means" in the accident—the constant breakdown of locomotives that demanded the use of the coal train and the fact that the train wasn't ready for service the next day. Although a breakdown launched the sequence of events, and workers had committed errors along the way,

Coddington singled out this "system" of working—the running of trains in both directions on one line—as the main cause of the disaster.[38]

The period of the Shields accident, the precise moment when the systems explanation emerged, was the high point of debates over the power of railways, whose interests they served, and how they could be held responsible for accidents under their control.[39] By the mid-1850s, the system accident formed a central plank in the campaign to reconfigure worker responsibility for material deterioration. As one report explained in 1856, although "inferior servants" were punishable for material failures, "many cases of negligence, for which these men have been punished, have been shown by the reports of the inspecting officers to be partly attributable to the defective system of working or to an insufficient establishment."[40] Collectively, the inspectors' accident reports supported a growing belief that almost all railway accidents were unnecessary, being preventable by systems of working that the government was unwilling to force on the railways, and which the companies themselves were too miserly to implement (figure 2.6).[41] By the mid-1860s, the time of Dickens's accident at Staplehurst, that form of explanation reached an apogee. And then it collapsed.

The Danger of Systems

By the 1860s, railways had become not only part of British metropolitan and imperial identity, but also one of its premier theaters of disaster.[42] Mining explosions, factory fires, building collapses, coach accidents, drownings, and shipwrecks were all part of a growing public fascination with accidental calamity, driven by newly politicized forms of danger.[43] Within this cosmology, though, railway accidents were distinct. Rather than hidden underground, or behind factory walls, or far out at sea, they took place in plain view outside towns, in densely packed cities, in fields, and in suburbs.[44] Newspapers photographed, illustrated, and reported on them extensively. Writers placed them at the core of fiction genres: railway novels and plays that helped cultivate a nervous traveling public. They helped generate a wider Victorian geography of peril that projected urban dangers into suburbs and countrysides, transforming risk from something one took on willingly through a business venture, for instance, into an inescapable element of modern industrial life.[45] Their impacts on inner lives would cast a long shadow over the late nineteenth century and beyond Britain, featuring in the intensely personal, microscale concerns of Karl Marx, William James, and Sigmund Freud. But their roots as public and psychic traumas that threatened everyone lay in the specific transformations of the middle decades of the nineteenth century and especially in Victorian Britain.[46]

From their earliest days, railways had also raised new and disturbing relationships between breakdowns and the inner lives of passengers. As Wolfgang Schivelbusch notes, "The breaking of a coach axle in the eighteenth century merely interrupted a slow and exceedingly bumpy trip on the highway; the breaking of a locomotive axle between Paris and Versailles in 1842 led to the first railroad catastrophe that caused a panic in Europe."[47] The enormous growth of the railway in the following decades had only amplified those effects. In 1845, Victorian Britons made thirty million railway journeys. Twenty years later, they made 239 million.[48] With that growth in the frequency and severity of accidents, medical officials and the popular press began identifying a shift in popular fear away from the external threats that railways posed—boiling, suffocation, colds, catarrhs, and consumption—and "toward the internal world of the human mind and body . . . a fear of insidious internal rather than catastrophic external disruption."[49]

Breakdowns formed a main focus of this shift, captured in an influential report that *The Lancet* medical journal serialized in 1862. The work of a scientific commission, it drew contributions from "the most able members of the Medical and Engineering Professions" and garnered so much public interest that the editors decided to publish it in collected form under the title "The Influence of Railway Travel on Public Health." One of its most influential claims was that railways produced a distinctly modern pathology of shocks that broke down both passengers and the material components of the railway.[50] The report described how the unremitting torsion on locomotive axles eventually snapped them. Rails that should have lasted thirty years wore out in three to six.[51] In humans, material shocks acted directly on the nerves, registering on two scales and producing two regimes of pathology. Continual small-scale concussions exposed existing, but often-hidden, debilities in people. Violent accidents occupied another regime of injury and produced a "violent concussion of the nervous centers" with "secondary effects" that were more serious than the immediate and obvious physical injury, as with Dickens at Staplehurst. These secondary effects emerged only later, sometimes evolving over time into mental degradation, paralysis, and blindness. Unlike other components of health, such as diet or hygiene, there were no obvious regimes of the self that could counteract them. In a society fascinated with visibility, surveillance, self-regulation, and self-responsibility, accidents and the breakdowns causing them stoked fears about latency, invisibility, and lack of control.[52]

But the role of breakdowns was not limited to their physical effects. Even when they failed to create an accident outright (e.g., by stopping a train in an unexpected location), they fed into Victorian nerves, creating the constant anticipation of collisions and calamities: "Every one knows how, if by chance a train stop at some unusual place, or if the pace be slackened, or the whistle sound its shrill alarm, a head is projected from nearly every window, and anxious eyes are on the look-out for signs of danger." Anticipating the concerns that would shake Dickens three years later, the report highlighted the dangers that this anticipation held for Victorians' inner lives:

> An often experienced condition of uneasiness, scarcely amounting to actual fear, which pervades the generality of travellers by rail. The possibility of collision is constantly present to such persons. . . . So, too, the frequent lateness of trains, and the bad time which they keep, are causes of anxiety. . . . The passenger is forced into subjective sources of mental activity; and where the tendency to excitement exists, this also, *quantum valeat*, must be esteemed an undesirable feature belonging to this manner of locomotion.[53]

By the time report was published in *The Lancet*, the Victorian popular press saw the inspectors' construct of "systems" amplifying these dangerous effects. The Staplehurst accident involving Dickens was a showcase for those concerns. The railway inspector who investigated it had characterized it as a system accident. Recall that the train had fallen through a gap where the tracks had been removed for repairs. A system of elaborate procedures involving fog signals, red flags, and an attentive signalman existed to prevent exactly this kind of event.[54] Had it been adhered to, the inspector Henry Rich claimed, the accident would have never happened. *The London Times* disagreed. Instead, it launched a broadside against the growing fetish for systems, echoing wider concerns about their catastrophic potential:

> The whole arrangement depended upon accuracy at every point. If the plate layer were accurate, and if the trains were punctual, and if the work were successful, and if the signals were properly placed, and if the day were not too bright nor yet too foggy, and if the engine driver were always looking out, and if the breaks [*sic*] were patent, the train might be safe; but *a failure in one of these elements* might render all the rest abortive. This is, indeed, the vice of modern railway management. The press of traffic is enormous, and its safety is made to depend upon

> a complex system, first of timing, and then of signals to counteract failures in punctuality. *Such a system requires constant and unfailing care.* . . . and then the Directors assume an air of the most virtuous innocence because their system looks well on paper.[55]

By the early 1860s, railway inspectors themselves had started expressing similar skepticism. In the period of *The Lancet* report and the Staplehurst accident, the most influential figure on these questions was Henry Whatley Tyler. He had been appointed a railway inspector in 1853 and would hold that position for twenty-four years, eventually rising to Chief Inspector of Railways in 1871. Tyler would help found the Science Museum in South Kensington, was partly responsible for the 1851 Great Exhibition (famous for its Crystal Palace), and would be elected to parliament in 1880. During his career, he would investigate London's 1867 cholera outbreak and determine mail transport routes to India. After retiring from the inspectorate in 1877 and being knighted in the same year, he became president of Canada's Grand Trunk Railway. In August 1861, following two major investigations of bridge collapses where cast iron supports collapsed under the weight of passing freight trains, he was called to investigate the Clayton Tunnel crash. Here, the combination of a failed automatic signal, a farrago of human miscommunication, a disregard for proper spacing of trains, the problematic design of telegraph switches, and severe overwork of employees created the worst accident in British railway history up to that point—twenty-three people killed and 176 injured.

Tyler cast doubt on whether even perfect local "systems of working," such as the one set up but disregarded at Staplehurst, could prevent accidents. He returned to the problem of people. "We are too well acquainted with the imperfection of human instrumentality to suppose that even if all the systems of working were rendered perfect, if all the requisite means and appliances were supplied; and if all failure of materials were provided against, there would not still be mistakes on the part of some, and neglect on the part of the officers and servants employed."[56] The problem, as Tyler saw it, was that any of these systems depended on larger labor conditions that made human workers unreliable: "It is impossible for men who are employed over periods varying from fifteen to thirty hours, to do justice either to themselves or to their employers . . . The class of men so employed is not so good as it would otherwise be. Discipline cannot be maintained among them. They become stupid and reckless. They make mistakes in their signals, or neglect to keep a good look-out from their engines. They have an additional inducement to resort to stimulants, and even an excuse for excess; and they return to duty after their hours of rest, scarcely more fit for their work than when they left it."[57]

With his testimony appearing in newspapers, Tyler put forward a vision of material perfection that could compensate for human fallibility. "The strength of material is now so far ascertained, the processes of manufacture have arrived at such a degree of perfection, and the proper principles of construction are so well understood, that there is practically hardly any risk of failure in the permanent way, the bridges, the engine boilers, or any of the machinery of a railway upon which safety depends, when proper trouble is taken to make them safe, and when they are not retained in use for too long a period, when due attention is paid to deterioration, decay, corrosion, and all the effects of wear and tear, and when a sufficient margin for safety is allowed."[58] Tyler's solution, encouraged by his successors, was to shift the responsibility for adhering to systems onto material arrangements—to embody systems in material technologies—and to shift the responsibility for breakdowns back onto the vigilance and character of Victorian workers. One final accident will illustrate that development and its consequences.

The Collapse of Systems

The train had departed Paddington Station three minutes late at 7:28 p.m. on August 30, 1878—a Friday—heading to Aylesbury. By the time it arrived at Princes Risborough, just before 10:00 p.m., the driver was anxious to be on his way. The train discharged its passengers. The guard blew his whistle and waved a green lantern, indicating "all clear." Seeing that the starting signal at the station had also been lowered, the driver, who had been a conductor for twelve years and employed by the company for thirty, started off. Fifty meters beyond the station, as the engine passed a crossing, the driver felt the engine violently twist, strike some wagons in the siding, and then flip over, pinning him and his fireman underneath. At the center of the resulting investigation was a diagram (figure 2.7). It would be used to dismiss the main signalman, Dick Holloway.

We have to imagine the diagram as a kind of "Marauder's Map," crawling with railway workers and late-summer travelers, freight cars being shunted onto sidings so that passenger trains could pass, darkness falling,

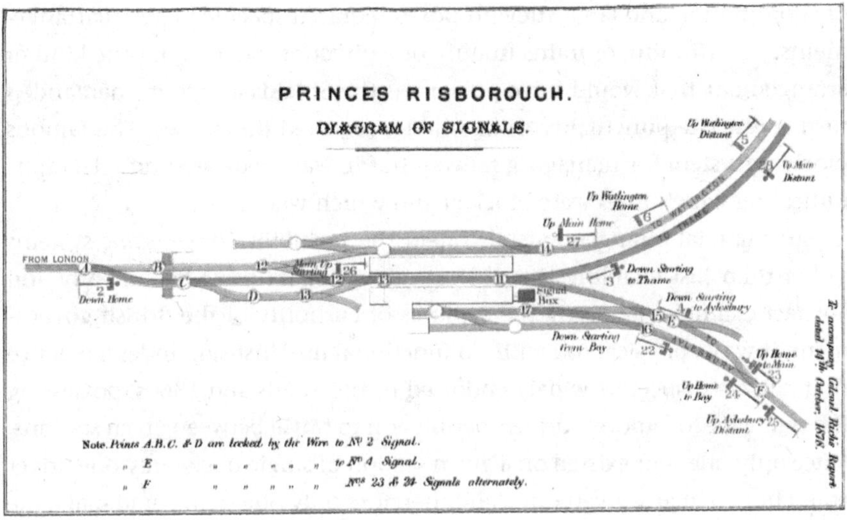

Figure 2.7
Diagram of signals for Princes Risborough. *Source: Parliamentary Papers* 62 (London: HM Stationery Office, 1879), 74.

dim lanterns waving to signal safety or danger. The signalman, Holloway, had arrived at work that morning at 9:00 a.m. and had found all the points, switches, and signals working properly. Near the end of a ten-hour shift, he had conducted a complicated shunting operation to put eight freight cars in a dead-end siding (extending beyond the points marked "E"). To do this, he had unlocked the points manually using a key kept in the signaling cabin. He had then gone on to unlock another set of points and returned the key to the cabin but had mistakenly left open the points at "E." Normally, an interlocking mechanism prevented the signal on the main line from being lowered unless the downline points were in their proper positions. But the mechanism had developed just enough play that it allowed the start signal to be lowered while the points were still out of alignment. When the train to Aylesbury passed junction E, the misaligned points deflected part of its engine into the siding, throwing it against the goods wagon and causing the collision.[59]

The accident caught Princes Risborough in the midst of a wave of material transformation slowly breaking across British railyards, junctions, and stations, but still incomplete in 1878. Since they began appealing to systems in the 1840s, railway inspectors understood that the abstract relationships of coordination and labor they proposed required specific material arrangements. The running of trains in only one direction on a track—the kind of arrangement that would have prevented the Shields accident—demanded turntables. The punctuality of timetables required timepieces. The famous blocking system for managing railway traffic was predicated on telegraphs indicating which lines were blocked and which were clear.

But material arrangements also held the possibility of *enforcing* systems rather than just enabling them. The "railway staff" provided an early and abstract example. Recalling the symbols of authority of the British government, it was a physical rod with no functional use. Instead, under a practice that railway inspectors widely endorsed in the 1850s and 1860s, possessing the staff gave locomotive drivers permission to travel between given stations. Since only one staff existed on a given section of single track, only one driver could hold it at any given time, and therefore only one train could make the journey through the section, preventing head-on collisions.[60]

More elaborate material arrangements, compelling certain actions and preventing others, became increasingly common between the 1850s and the 1870s. The discretion of railway officials—stationmasters and other officers of the company—formed one of their primary targets.[61] But the expanded systems also increasingly enveloped the work of enginemen, guards, and switchmen. As a site of maximum discretion and maximum danger, traffic signals were one of the main sites for this materialization.

Starting in the late 1850s, Board of Trade inspectors such as Tyler began recommending that proper systems of work should be embedded within increasingly complex automatic signaling mechanisms. One report in 1858, for instance, called for signals and points (the section of tracks that switched trains onto other lines) to work together "and be so arranged as to prevent the signalman making a mistake that will be dangerous to the safety of passing trains."[62] The recommendation drew from John Saxby's interlocking mechanism, introduced two years before and designed precisely to deal with the problem of accidents like the one that would happen at Princes Risborough—a disagreement between signals and points.[63] In 1873, driven by the calls of the Railway Inspectorate through the Board of Trade, a bill that would have compelled all railway companies to adopt the "block and interlocking" system was introduced to Parliament but was ultimately put aside in favor of monitoring the transitions already underway, of which Princes Risborough formed one part.[64]

Monitoring these new material arrangements for signs of breakdown, though, was not as easy as engineers such as Tyler predicted. Although the new materialized systems seemed to relieve workers of the need for constant vigilance and rule-following, in practice, they placed the proper operation of an expanding coterie of machines and devices under the watch of fewer and fewer staff. The new London Bridge signaling system (figure 2.8), exhibited at the 1880 exhibition in Paris and a showcase of the interlocking system of the time, featured 280 levers in a single raised cabin, perched like a panopticon over the tracks. It was worked by just four signalmen at any given time, rotating on eight-hour shifts.[65] The unlikely alliance it created between the Railway Inspectorate, who saw it as a materialization of the systems they advocated, and the railway companies, who saw reduced labor costs, increased traffic capacity and profits, shaped the view of who or what should monitor the new arrangements for possibilities of failure.

The irony of these material systems lay in how they made workers responsible for sensing the potential failures of mechanisms designed to regulate their own actions. Throughout the 1870s and 1880s, as their adoption expanded, so too did ancillary mechanisms to enforce that monitoring and make it part of the accounting that workers gave of themselves

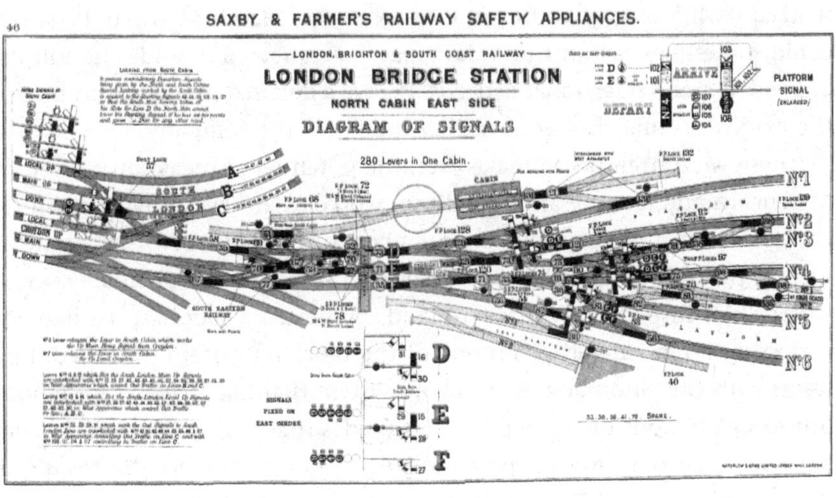

Figure 2.8
London Bridge Station diagram of signals. The diagram notes 280 levers in the cabin located just below and to the right of the text. *Source:* Saxby & Farmer's Railway Safety Appliances, January 1889, 4th ed., 46.

as laborers. The aptly named Punishment Books of the period illustrated the techniques through which workers identified their own value with their ability to notice material breakdowns.[66] In a practice that would run into the 1940s, companies kept a ledger entry for each engineman they employed. Sometimes stretching over four decades for a single worker, the detailed records listed both good deeds and offenses, often labeled "misdemeanors." The origin of the term "misdemeanor" is significant here. Even in the mid-nineteenth century, "demeanor" pointed to a form of conduct, a way of acting, or a manner of living, and its links to spirituality and religion often tied it to the notion of rewards or punishments.[67] More than just misconduct, "misdemeanor" signaled ways of acting that were suspect.

The ledgers went beyond simply listing good deeds and misdemeanors. They administered monetary rewards and fines, respectively, for each. Disobeying signals or running past stations might incur a five-pence fine. But notably, the ledgers included in these lists of virtues and misdeeds an engineman's awareness of material flaws they had encountered and how they had responded to them. The language of the entries is particularly telling. Enginemen with the South Eastern Company, for instance, were rewarded not simply for reporting flaws but rather for "observing" them. Conversely, Henry Barnes was fined for "not seeing" that a brass rod had worked itself loose before he left Canterbury on March 31, 1876. Penalties could sometimes amount to a day's pay.[68] Arranged alongside the full list of good acts and questionable deeds, it gave a detailed moral accounting of the worker's comportment over their time with the company.

These were attempts to make even the potential of breakdown a source of concern and anxiety among railway workers. Not even these measures, though, were enough to put the members of the Railway Inspectorate at ease. Reprising the question of signals in 1881, Tyler's immediate successor as chief inspector, William Yolland, presented a report to parliament endorsing a new material "union" of the blocking system for scheduling trains with the automatic interlocking system that had spread throughout British rail stations over the previous decades (figure 2.9). The new system allowed "each train to exercise a most important *control over the signalmen* in the signal boxes." The mechanism not only signaled which lines were clear and which were occupied but also locked the levers for the points so that signalmen could not activate the switches unless the telegraphic system indicated a clear line. Yolland explained how the new system was not

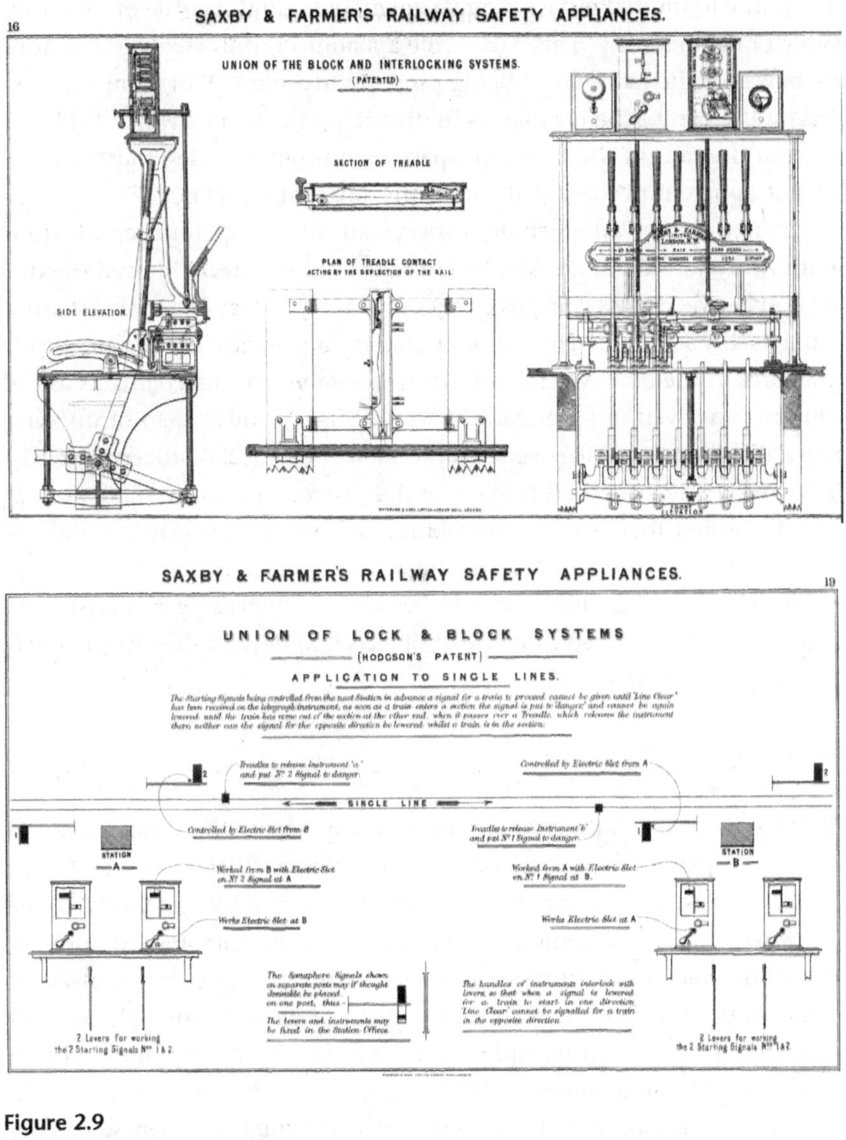

Figure 2.9
Saxby's patent for the block and interlocking system. *Source:* Saxby & Farmer's Railway Safety Appliances, January 1889, 4th ed., 16 and 19.

only prone to limited modes of breakdown, but had the virtue of reporting its own malfunctions: "This part of the apparatus is only likely to fail from the power of the battery not being properly maintained, or from carelessness on the part of the platelayers in disturbing the treadle which is placed between the rails in the four-foot space, and if out of order, must at once become apparent to the signalman in the adjacent signal box."[69]

Reported in the Parliamentary Papers alongside telegraphic reports from South Africa and the Boer War, Yolland's description freely blurred the distinction between the system as a way of working and its material instantiation in Saxby's product. His vision of human carelessness as the sole culprit of failures in the new system, and the dependability of mechanisms as the solution, was even more extraordinary because of the farrago of ongoing material failures his office oversaw and catalogued—1,238 broken tires, 514 broken axles, 12 collapsed bridges, and 446 broken rails in 1880 alone.[70] And it pointed to the expanding kinds of workers—specialized maintenance staff, fabricators, operators—for whom breakdowns in the material arrangements of the railway would become an increasing responsibility, along with the expanded universe of objects and devices they would oversee in the century to come.

Conclusion

The growth of the railways naturalized many aspects of Victorian social relations as part of the experience of modernity.[71] Whereas accidents had previously been freak occurrences, by the 1860s, railway systems were the sites where impossible conjunctions of imperfect schemes, flawed humans, and failing machines routinely came together to threaten a newly nervous public. Against our first intuitions, the failures of those technologies weren't simply suspensions or interruptions in the formation of responsible liberal selves and industrial subjects. Rather, they were moments for Victorians to assert critical aspects of their social order by expressing their sensitivity to them: how breakdowns interrupted their frantic lives and free movement, frayed their nerves, or registered in their working worlds and laboring selves. Those assertions worked differently across the Victorian social order. Railway inspectors saw them as sites where they could reconfigure pressing, anxious questions about material degradation and responsibility amidst the challenges of distributed causality. The anxiety that middle-class travelers felt wove breakdowns into questions of self-responsibility, self-governance, and even freedom. The Punishment Books catalogued how breakdowns were turned into a moment for workers to assert themselves as good employees and, by extension, good subjects of the liberal economic order. The larger materialization of Victorian systems—sanitation, punishment, taxation—increasingly repeated that experience in British factories, households, and imperial infrastructures. And they inspired ever-expanding materialized systems designed not only to keep Victorians acting responsibly but also to make them identify with material breakdowns through techniques of industrial capitalism that echo to this day. We'll turn to one of the most prominent of those echoes, the Gantt chart, in chapter 4. But for now, we turn to another kind of breakdown.

3 The Breakdown

Before breakdowns touched machines, they moved people.[1]

You descended into Almack's as if going underground. Beneath a carpenter's shop at 67 Orange Street in the Five Points, after a trek through mud and narrowing darkness, you approached the door of the notorious dance hall. Beyond it, a shilling poorer, a raucous cross-section of antebellum New York City gathered most nights. The Black owner, Pete Williams, was a wealthy and successful impresario in a time and place that made Black proprietors scarce. A patron of the nearby theater at Astor Place, he was, according to the era's best-selling city guide, a first-rate critic.[2] Williams had owned a string of places on Orange Street before hitting on Almack's. But it was here, on the night of March 4, 1842, amid the booze and din, a decade before Staplehurst, that Charles Dickens witnessed a dance—a "regular breakdown" that would celebrate a Black cultural form and provide the most memorable passage of Dickens's *American Notes*.[3] Polite Gotham society never forgave him.

The things that went on in Almack's made it detestable to New York City's establishment and invisible to much of history.[4] At the time, though, its interracial mixing and illicit indulgences made it all too visible to the city's respectable elements. Almack's was Manhattan's open secret. Its clientele mirrored the social admixture of the Five Points. Known as the nation's most prominent slum, the neighborhood was the product of extreme social forces that went beyond mere economics.[5] In the seventeenth century, it was known as Fresh Pond and was the colony's first designated Black community. In 1741, when it was still forested hills, this had been the site of an alleged conspiracy between enslaved Blacks and poor whites to raze New York City and overthrow the government. Anti-abolitionist rioters, fearing Black ascendancy and miscegenation, tore through the neighborhood in 1834, attacking abolitionists and ransacking Black establishments and homes. By the 1850s, the area was mostly home to newly emancipated Black New Yorkers and newly arrived ethnic Irish, who competed with the established Black community for work. It was also a tourist area for men from other parts of the city—part of the flood of single white men who arrived in New York starting in the 1820s, searching for work, and who consumed the spectacles and behavior that more respectable audiences decried, while they mixed with the rough worlds of the urban working classes.[6] As sites of interracial socializing, places like Almack's had many enemies.

Before the word "breakdown" ever named a mechanical event, it denoted a dance performed in places like this and many others. The hidden transcript of the dance symbolized a moment of Black spectacle and inversion.[7] It had originated as a celebration among the slave cultures of antebellum America, performed during a pause in harvest labor. As such, it also celebrated a break in the work of machines—cotton gins, threshing machines, corn mills. It was the moment when machines stopped and people rested and celebrated. Sometimes referring to a form of dance, sometimes to an event that gathered several energetic dances in a single spectacle, it was performed in the North as part of the slave holidays "Pause" or "Pinkster."[8] The origins of its name are a mystery. Some said it honored the rhythmic stomping that broke down floorboards; others that it borrowed its name from the activities it accompanied—the breaking down of crops during the fall harvest. White performers stole the dance and began performing it in blackface.[9] Their white audiences consumed its spectacle alongside an industry of racial caricatures. But for its Black performers and audiences, the breakdown also meant something else.

The performance that Dickens witnessed was the work and art of William Henry Lane, probably born in New York City in 1824 and better known under his stage name, Master Juba. Lane would become a star of the popular press and the circum-Atlantic entertainment circuit, breaking free of Almack's and its associated acts and eventually performing in England for Queen Victoria.[10] The year before his performance, he had worked for the impresario P. T. Barnum, who tried to pass him off as a blackface minstrel to patrons at Vauxhall Gardens. The following year, undeterred and likely emboldened, Barnum advertised Lane as the white Irish minstrel performer

John Diamond, a rival who danced breakdowns in blackface and sometimes impersonated Lane himself. The resulting competition produced a direct confrontation between "these Two Celebrated Breakdown Dancers."[11] Lane's performance became a triumph and a revenge against impersonation and fraud. He would use the breakdown to assert his own identity against competitors and imposters, imitating his rivals' inferior styles before "imitating himself" to show how his style differed completely.[12]

Through his performance, Lane connected breakdowns as dance to a nascent genealogy of Blackness and machines. The art critic Marian Hannah Winter would see him as the single most influential performer of the nineteenth century, whose performances were a progenitor of jazz and tap dance—the paradigmatic cultural expressions of the Machine Age.[13] Lane took his stage name from the Juba dance, a predecessor of the Charleston and a competitive dance of skill, brought to America from the Kingdom of Kongo and performed alongside the breakdown in Congo Square, New Orleans. This French Quarter open space, where slaves were allowed to gather on Sundays, creating a vibrant admixture of African, creole, and "American" music and dance, would produce the other great seedbed of jazz.[14] In its energy, carnivalism, promiscuity, and hedonism, jazz, like Almack's and Congo Square and Lane's breakdown, would represent a threat to racial purity. It would also provide the prime medium for Black America's assimilation of the machine into American popular culture.[15] As the Black community of the Five Points created a new future for itself uptown in the early twentieth century, they would bring those associations to the clubs and speakeasies of Harlem, whose syncopated rhythms evoked the steam engine and its fugitive geographies.

* * *

For Black steamship workers, free and enslaved, plying the Mississippi in the 1840s, retracing the journey of the "breakdown" from Southern plantations to Northern cities and back again, mechanical breakdowns took specific and dangerous forms. Steam was a temperamental technology in the early decades of the nineteenth century. When it worked, it followed a logic of narrowly contained violence: intense heat and immense pressure confined in imperfect steel vessels that became shrapnel when they exploded. Between 1816 and 1848 in the United States, 233 steamboat explosions killed 2,563 people and injured 2,097.[16] The engine's mixture of iron, water, fire, and corrosive ash, working in the humid heat and muddy waters of the South, was a recipe for corrosion, decay, leakage, and rupture.

In the years surrounding the Civil War, Black Americans used this technology to create a "rival geography" operating beyond the surveillance of the plantation system.[17] The river network had enriched many Southern enslavers. It had also created a social and commercial space beyond their control, where free Black people spread abolitionist ideas and abetted slave escape. Deckhands stowing freight, roustabouts loading and unloading it from piers and levees, chambermaids tidying and ordering the boats' parlors and cabins, firemen stoking furnaces—together they formed a mobile community of two thousand free Blacks and three thousand leased-out slaves that lived between the ships and the port communities they visited.[18] They turned steam into a Black fugitive technology, creating a breach in the systems of Southern racial control and shaping a geography of both hope and peril. The dangers of steam explosions sometimes abetted their escapes, hiding the flight of fugitives that white overseers took for dead.[19] For escaped slaves such as William Wells Brown—the abolitionist, novelist, playwright, and historian who fled in 1834 while working on a steamboat on the Ohio River—breakdowns were filled with the anxiety of potential discovery and capture, a delay in dangerous lands, even when they didn't extinguish lives outright.[20] Henry Bibb, on his way to freedom aboard a steamship bound for Cincinnati in 1837, couldn't sleep for fear of a boiler explosion: "Every time the boat landed and blowed off steam, I was afraid the boilers had bursted and we should all be killed."[21]

Watching for these calamities was the work of Black steam men working below the pilothouse and the posh saloon of the steamer's main gallery, areas forbidden to Black passengers, no matter how free or wealthy. In front of incandescent furnace openings arranged in a row on the main deck, they loaded wood every fifteen minutes, stirred embers, and cleaned out the boilers during stops.[22] They worked separately from the white "native-born" and immigrant laborers—mainly Irish and Germans—but formed part of a growing class of Black "mechanics" in the antebellum United States, whose rise and prominence threatened the project of Southern secession and the racist theories that supported it.[23]

On his travels through the seaboard slave states, Frederick Law Olmsted, the future architect of Central Park, would report on the enslaved engineers who built and maintained the plantation cotton gins and repaired its steam engines.[24] In *The Negro Artisan*, W. E. B. Du Bois pointed to the antebellum Southern railroads, whose entire crews, including engineers and fireman, were made up of enslaved men, extending the riverine system and bringing it within reach of landlocked plantations.[25] Tending to these steam machines, fixing them when things went wrong, formed a critical locus for a set of practices, experiences, and understandings of Black male selfhood.

The threat of steamboat explosions formed one more danger against which the steamboat workers shaped a form of life they called "rascality." Rascals were men who lived by their wits, combining toughness and strength with an adaptability and a flamboyance that evoked the dandy.[26] Their actions sometimes sought personal gain at the expense of other working-class people, but they also intersected with forms of Black resistance that targeted the region's elite.[27] They were agents in a network of illicit trade, acting as fences for possessions that enslaved laborers had stolen from their masters. They themselves stole from white wealth—plantation owners, bankers, merchants—while occasionally helping to organize slave escapes.[28]

Rascality named this mix of danger, mobility, cleverness, conspiracy, and relative status. For centuries and across oceans, that term had been used to describe the dishonest rabble of the lower classes. White antebellum Americans used it to characterize what they saw as the natural tendency of African-descended people toward moral corruption.[29] The steam men took the term of abuse from their white overseers and turned it into a point of pride.

In 1851, the white Louisiana physician Samuel Cartwright would transform this expression of Black liberty into a set of pathologies. Amid the flourishing tradition of Black mechanical expertise in the antebellum American South, Cartwright published a paper in the *New Orleans Medical and Surgical Journal*. Entitled "Report on the Disease and Physical Peculiarities of the Negro Race," it relayed the findings of a four-person all-white committee appointed by the Medical Association of Louisiana and chaired by Cartwright. After listing physical attributes that supposedly proved Black inferiority and justified enslavement, Cartwright coined two diseases that afflicted Black people exclusively. The first was *drapetomania*, "or the disease causing slaves to run away." The second was *dysaesthesia aethiopia*. "To narrate its symptoms and effects," Cartwright claimed, "would be to write a history of the ruins and dilapidation of Hayti." White overseers allegedly mistook the effects of the disease for deliberate action and called it "rascality." Cartwright instead argued it was "the natural offspring of negro liberty," afflicting virtually every free Black person, making them not only insensible to pain but also insensitive to the spoiling of crops, the abuse of animals, and the *breaking of tools.*[30] Cartwright drew on a long history of racist science claiming Black insensibility to physical suffering, and extended that insensitivity to the breaking of technologies. The report concluded by suggesting the disease—which produced the inability to be disturbed by broken machines—refuted the equality of the races, invalidated the universality of the Declaration of Independence, and made Black Americans unworthy of liberty and republican institutions.[31]

* * *

Beyond the reach of the Mississippi and its watershed lay the Low Country of South Carolina and the Sea Islands of Georgia. This is where Gullah culture and Black rice first emerged from the knowledge and techniques that Africans preserved over the Middle Passage.[32] Throughout the 1840s and 1850s, this was also a geography filled with engines: locomotives running from the Georgia coast to the interior, steamboats on the Savannah River, ocean steamers shuttling cotton from the jetties of Charleston destined for New York and Liverpool, and the cotton engines ("gins") that enslaved workers used to separate the most valuable cotton in the world—Sea Island cotton—from its seed.

Those seeds had originally come from Caribbean plantations. The successful slave revolution in Haiti had scattered French Caribbean planters across the Antilles and into the Southern United States, including the Low Country, where they occasionally converted rice production to cotton when prices soared.[33] The long-strand cotton crop thrived only in this swampy area near the coast, outside the reach of the river systems and their deltas further inland.

That watershed marked not only fluvial geographies but also hard-won concessions that slaves had secured, mobilizing precious crop knowledge in cunning negotiations with white overseers that shaped antebellum Southern society and daily life.[34] One of these was the task system. Through it, slaves had set limits on the work any enslaver could expect in the Low Country, allowing them to finish their work before dark, leaving them free to determine the pace of their work during the day, and earning time for cultivating their own gardens or tending to family and companionship in the evenings. The task system also required less oversight and fewer rations of food, keeping enslavers themselves out of the malarial swamps.[35]

By working slowly and with deliberate inefficiency, by deceiving, playing dumb, breaking tools, stealing, and pilfering, enslaved people protected this form of work against the more exploitative gang-work systems that characterized slavery in the Deep South.[36] Under constant threat and coercion, they broke down their overseers plans through secret methods, as a way to contest power.[37] When they were sold across the divide between inland and coast, into the expanding cotton empires and the even more brutal gang-based work further inland, they experienced the move as a transformation of inner worlds—a horrible moment in their self-construction as they were pushed into new systems of power and labor extraction that, in other guises, would soon inspire the techniques of American capitalism and industry.[38]

They also felt that transition in the machines they tended. The Sea Island cotton cultivated in the Low Country of South Carolina was prized for its long fibers. It couldn't be processed with Whitney's famous gin or its successors. Workers in the Low Country instead used older-style roller gins. For the women and men operating them, this was a bodily affair. The roller gin was meant to replicate the careful work of hands pinching seeds from cotton fiber. Previous hand gins had required workers to turn with one hand and feed cotton with the other, or to work in twos. Here, ginners instead sat at the gin and powered it using a treadle at their feet. The treadle powered flywheels, attached to rollers. Resting her forearms on a small platform in front of the roller, the ginner fed the raw cotton fiber between the rollers, which separated the seed from the fiber.[39] The arrangement made the entire machine an extension of the body.

In doing so, it demanded things of enslaved bodies and minds. In the late eighteenth century, enslaved people performed most mechanical occupations in the Lower South. Claims that they could not be artisans or machine operators were inventions of antebellum proslavery campaigns.[40] And they were belied by escaped slave notices describing the craft skills and expertise of fugitives; and by the vigilance and attention machine operators deployed to make machines work and to keep them working. Echoing a common refrain about the technological ineptitude of Black laborers, one planter claimed that a new roller-type gin "is the most simple in its construction . . . and on that account, extremely well adapted to the use of negroes."[41] He went on to attribute its durability to its use "of the most durable materials, being principally cast steel." But his description of the new device told another story. Its durability required constant intervention and skill. Operators had to regulate the distance between its rollers, made of cast steel and Northern hickory, using a pair of thumb screws. Its motion was regulated using an iron balance wheel requiring adjustment. The friction wheels needed regular oiling so they didn't overheat and seize or spark and cause fires. "Should the gin at any time be clogged by the accidental introduction of a foreign substance, or should the upper roller require to be renewed, or removed for any other purpose," enslaved operators would take it apart and reassemble it.[42]

To work, the machines required something else: a disposition. "Active, ambitious" operators spun the machines' rollers at 250 revolutions per minute.[43] Earlier versions of the device produced enormous strain on the back. The pressure on the forearms and the fatigue of the fingers and the legs required occasional breaks "to get breath."[44] The twelve pounds of lead weights on the flywheels spun with enormous momentum, requiring strength and stamina to run them. They were "apt to rupture negroes and always injures them, if continued for any length of time," sometimes leading to hernias or miscarriages.[45] And then, there were the punishments when things broke down.

The suspicion was that breakdowns were acts of *sabotage*. But Black gin operators knew much more than this. Breakage was something other than what white overseers and technologists understood. More than a century before academics would theorize that, for machines to work, people had to adapt to their affordances, molding their selves to the technology's needs, enslaved peoples had turned that observation on its head.[46] Overt resistance and sabotage were unnecessary. The fabrications of racist science concealed a deeper and more abiding truth about these and all technologies. Their operation always rested on a knife-edge, like the one separating cottonseed from lint. It didn't require spectacles of destruction and crafty subversions. Molding a certain kind of self—calibrating and regulating one's body, holding it in constant tension—was what kept these machines running. For machines to break, it was enough to *rest*.[47]

* * *

Maybe you would have liked to stay longer in Almack's, to feel the breakdown dance move you as well. Or to see how rascality could celebrate rebellion and refusal. Or to hear firsthand the hopes and dreams of the men and women working the gins. At some point, though, knowledge must fail us. And we are left with an ethic: to bear witness to other ways of being—and to see in their histories echoes of the breakdowns that vanish so quickly from our own.

4 Birth of a Notation: Charting Human and Machine Failures in the Progressive Era

> It is a peculiar sensation, this double-consciousness, this sense of always looking at one's self through the eyes of others, of measuring one's soul by the tape of a world that looks on in amused contempt and pity.
>
> —W. E. B. Du Bois

Autumn, New York. Headquarters of the "New Machine." Sometime in late 1919, a man set out to save modern civilization with a chart. The son of Maryland slave owners and a former protégé of the efficiency wizard Frederick Winslow Taylor, he had helped develop the central apparatus of Taylorism—time-study techniques, wage schemes, and workplace reforms that would transform laboring lives and spark the resistance of workers across American industry. He would go on to become one of the most influential of all Progressive Era efficiency consultants, advising Bethlehem Steel, Westinghouse Electric, Remington Typewriter, Canadian Pacific Railway (CPR), and the US Navy. It was in his living room in 1910 that future Supreme Court Justice Louis Brandeis coined the term "scientific management" to describe the powerful movement taking shape. His graphs first charted the work of women spoolers and male weavers in a factory in Passaic, New Jersey, in 1908. He had elaborated and extended their techniques to analyze weapons manufacturing and ship production for the US Bureau of Ordnance during World War I. Shaken by that recent cataclysm, he now proposed the final iteration of his graphic technology: a chart that, he believed, would secure the regular and efficient functioning of the American industrial system by graphing the interrelated failures of people and machines.

The Gantt chart would become one of the most influential and pervasive tools of modern management consulting—a fixture of twenty-first-century business plans and production schedules, and a main selling feature in twenty-first-century productivity software and project management. The chart's origins, however, lay deep in a racial and civilizational vision rooted in the Protestant ethics of Progressive Era America. Histories of the Gantt chart have concealed those origins.[1] Far from a mundane instrument of detached technical efficiency, Gantt's chart formed part of a self-proclaimed heroic struggle to create a racially purified American industrial polity, renewed not so much through the virtue of efficiency, as most histories of Gantt claim, as through the elimination of idleness.[2] The bourgeois ordinariness of the graph's ruled lines betrayed an almost mystical belief in the power of paper technologies to transform "inner lives" and shape outer worlds. But only for the elect. The story of the Gantt chart is a story about the formation of the white technological self in turn-of-the-century America and the acts of violence committed in order to protect it.

At the core of that story are the breakdowns of machines. The "paper technologies" of scientific management—charts, diagrams, instruction cards, questionnaires—were the media through which Machine Age Americans accounted for their labor, their moral character, and their time.[3] They were also the instruments through which breaking machines were transformed into both a civilizational problem and a problem of the laboring self. In the rough game of American industrial capitalism, in an age obsessed with both the secrets of phenomenal success and the moral conditions of individual calamity, notions of failure crisscrossed the human and the machinic. Gantt's charts represented a final codification of a long-standing anxiety within scientific management about how breaking machines should be tied to the interior lives of workers. In early twentieth-century America, the charts did much more than measure the efficiency of humans and machines "at work." By the dawn of the Jazz Age, they revealed the intricate cultural linkages and hidden dangers that connected two kinds of failure for white Americans: failure as a condition of industrial machinery, as a breakdown, and failure as a kind of person.[4] Linking these two conditions through his graphs, Gantt came to see his charts as the principal instrument in two sweeping and interrelated projects: eliminating idleness as a sin that afflicted both broken industrial machines and individual white workers, and keeping Black workers out of the Northern industrial sites under the Great Migration.

The chapter begins by exploring scientific management's roots in the regeneration projects of the Gilded Age and the Progressive Era. Although it is often cast as a cold and bloodless system for creating efficiency, scientific management was part of Progressive Era projects that linked a reborn American society with the transformations of the soul. Springing from Protestant notions of conversion and predestination, those projects took many forms: imperial war, patent medicines, and the "gospel of efficiency" that underpinned Gantt's work.[5] To understand how those larger impulses were transformed into seemingly objective instruments of industrial management, the chapter turns to Gantt's paper technologies. As an efficiency consultant, Gantt came to believe that working selves and the machines they tended were linked, not through the technical and personal virtue of efficiency but rather through the vice of idleness—an inclination drawn from both American cultural history and his family's personal demons. Breakdowns and the idleness they produced were central to this scheme. The son of Maryland slave owners financially ruined by emancipation and physically incapable of industrial labor, Gantt saw idleness as both a machinic and a personal failure, and his graphic technology as a site of personal redemption and transformation. The closing section of the chapter examines how, by the end of World War I, this tool of transformation formed an instrument of racial containment. Deployed within the white-protected industrial concerns of northeastern workspaces, the charts codified the supremacy of white workers, hid the ongoing nature of social domination, and recast it under the guise of progress.[6] Tracing those concerns across steel mills, metals shops, and textile factories, the chapter explores how Progressive Era efficiency experts, factory managers, and workers came to understand breakdowns as a problem of the self—events that revealed the *kinds* of people who belonged in the theaters of American industry and the civilizational mission of turn-of-the-century America. The focus on Gantt here is not an effort to memorialize an author of technological racism. It is meant to recast the dark "underside" of scientific management as part of the active and productive racial fantasies that also underpinned and propelled it.[7]

The Selves of Scientific Management

In October 1910, the nation's leading efficiency experts gathered in an apartment at 120 West 57th Street in Midtown Manhattan. Louis Brandeis, a fifty-three-year-old Boston lawyer and soon-to-be Supreme Court justice, had invited them there. Brandeis was preparing to argue what would come to be known as the Eastern Rates Case. Railways wanted to increase their freight charges. Brandeis felt the rates were arbitrary and that railways could avoid a hike through increased efficiency.[8] He was looking for star witnesses, and he found them in his fellow guests: Frank Gilbreth, co-creator, with his wife Lillian, of micromotion studies; Henry Sheel; Robert Kent; and Henry Laurence Gantt, whose living room they were sitting in. Gantt was considered "the surest result-getter" of all the efficiency consultants.[9] The high priest of the movement, Frederick Winslow Taylor, was absent. But much of Taylor's own history and insights were present in Gantt, whose work under Taylor at Midvale and Bethlehem Steel had helped develop the central elements of Taylorism. Their movement only lacked a name to match its energetic purpose. Brandeis suggested calling it "scientific management."

Scientific management grew out of the Gilded Age and Progressive Era's emphasis on purification, regeneration, and rebirth—an emphasis that Jackson Lears has traced to the foundations of American personal and political life between the Civil War and World War I.[10] Like the antebellum movements for temperance and abolition, the era's regenerative projects—imperial expansion, patent medicines, prohibition—promised both personal and national salvation. Rooted in the ongoing psychic traumas of the Civil War and in a Protestant ethic that saw conversion as the central experience of salvation, those projects tapped American mythologies of new beginnings and reinventions of the self. There and elsewhere, Progressive dreams of national rebirth were founded on visions of a renewed white supremacy and internal racial terror that purified electoral politics by disenfranchising Blacks, codifying segregation through Jim Crow laws, and reasserting white identity as "American."[11]

The Progressive "gospel of efficiency" promised this combined social and personal transformation in its very name. Its claims to solve the nation's pressing concerns—gender inequality, immigration, industrialization, urbanization, corruption—drew on Protestant-inspired jeremiads tying the moral degradation of individuals to the moral fate of the nation.[12] Other key elements of scientific management similarly turned on this Protestant ethic. The movement's timesheets and task cards were pitched as instruments of personal bookkeeping and constant surveillance that would halt degradation and drive personal revival. The movement's promise of releasing the potential productivity of workers was of a piece with the broader conversion experiences that would liberate the productive potential of the saved soul.

More than a system for *making* efficient people, scientific management was initially a technology for *revealing* them. Frederick Winslow Taylor had built his foundational understanding of scientific management on a version of predestination: "Men are not born equal, and any attempt to make them so is contrary to nature's law and will fail."[13] People were instead born efficient or inefficient, winners or losers. Like Protestant journal keeping and self-examination, rather than making people into the select, Taylorism sought to distinguish the elect from the reprobate, drawing on a deep well of Protestant practices of self-examination to detect the workings of grace in the form of efficiency.[14] Through them, individuals revealed their membership among the chosen. The reborn would, in turn, remake society along more democratic or stable or equitable lines.[15]

The problem facing the sitters in Gantt's living room in 1910 was how to make this innate personal virtue visible. In the nineteenth-century ideal of Protestant America, such a mechanism would have been unnecessary. The ideal self was sincere and transparent. Inner life (especially grace) and outward appearances corresponded perfectly.[16] And few things revealed inner virtue as much as hard work and the success it was supposed to bring. In the cut and thrust of the Progressive Era, though, things were less obvious. How did one distinguish between the confidence man and the millionaire? (Taylor himself embodied the shaded line between "titan of industry" and "con man" better than most.) Credit-rating agencies, bankruptcy courts, and charity bureaus developed techniques to classify individuals as actual and potential successes and failures, calculating their social worth as individuals.[17] The establishment of Ellis Island as the immigration threshold to the United States after 1890 embodied the Progressive fascination with seeing into the souls of men, weeding out "all people of bad character, the incompetent, the lazy, the vicious, the physically unfit, defective, or degenerate," and determining their "fitness for citizenship," as Theodore Roosevelt would put it.[18] The tools of scientific management would add to this collection of instruments designed to reveal true selves.

Although efficiency appeared to cut across the deep racial divisions of the Progressive Era, in practice individual efficiency defaulted to whiteness. Black efficiency was framed around the needs and aspirations of the entire race and its institutions and always cast as a deficiency. On March 4, 1910, the President of the United States, William Howard Taft, spoke at Wilberforce University about the "crying need of trained [Black] leaders to direct the energies and activities of the race along lines promotive of the greatest efficiency."[19] Many Black leaders and commentators adopted these logics. Black institutions such as Fisk University were hailed for their efficiency. Public schools, churches, and community groups were exhorted to develop it. In May 1913, the president of the New York National Association for the Advancement of Colored Peoples (NAACP) established an "efficiency prize," not only to reward the "decent Negro" but also to "call the attention of the country to the fact that we produce not criminals and idlers, as many are prone to think, but men of high character and honorable ideals."[20] This racialized understanding of efficiency as a property of white individuals and an elusive goal of the Black race would shape scientific management's earliest techniques of revelation.

Of all the scientific management consultants, Taylor elaborated the purported techniques for revealing efficiency earlier, more fully, and more ruthlessly than anyone else.[21] Any shop floor, Taylor believed, contained two types of people: average men and "first-class men." At Midvale Steel Works, near Germantown, Pennsylvania he set out to demonstrate it. Since its opening in 1867, Midvale's primary business was manufacturing tires and axles for railway cars. By the late 1870s, it had begun receiving navy contracts for artillery, steel beams, and girders destined for the Brooklyn Bridge. While overseeing production at Midvale in 1895, Taylor introduced his famous differential piece rate, where workers meeting their production quota were paid a higher piece rate than those who failed.[22] His intention was to weed out slower workers, forcing them to quit. That same year, he complemented the scheme by hiring Black workers against prevailing rules and racist union practices that largely kept them out of industrial employment in the North.

Taylor's initiative was hardly inspired by egalitarian commitments. He aimed to use racial identities to dissolve the "clannish spirit" of ethnic cliques and regional loyalties—a project squarely within the traditions of racial capitalism.[23] Machines were central here. As Taylor noted, steel manufacturing was organized around large machine tools run by work gangs (figure 4.1) under the direction of a foreman, who usually surrounded himself with workers of his own ethnicity. Taylor envisioned the work gangs as pseudo-crime rings, plotting in foreign languages to hold down production through "soldering"—deliberately and collectively slowing their pace, and thereby concealing individual capacities under the cloak of the collective. In using Black workers, Taylor believed he could break the solidarity of the

Figure 4.1
African American workers at Midvale Steel and Ordinance Company's plant in Coatesville, August 5, 1921. Hagley Museum and Library. Box 10, folder 6–7, accession 1969-082, identifier 1969-082-I.-B.

gangs, increase individual competition, and reveal the true capacities of white workers. W. E. B. Du Bois would detail the experiment in Taylor's own words: "So we began bringing in Negroes and placing them on different gangs, and at the same time we distributed the other nationalities. Now our gangs have, say, one Negro, one or two Americans, an Englishman, etc."[24] Although Taylor claimed the system ran smoothly and increased output, by the late 1890s, Midvale had abandoned the experiment and created a strict racial segregation in its place that lasted until World War I.[25]

"One Negro, one or two Americans." Taylor's naturalization of whiteness as Americanness expressed the innumerable snubs and slights that shaped the Black double consciousness that Du Bois gave voice to. [26] For Taylor, Blackness—as non-American and non-white—was a tool to intervene in and disrupt existing forms of industrial life. More generally, the episode at

Midvale pointed indirectly to the unspoken common history that bound scientific management and its early practitioners to American slavery.[27]

The language of slavery pervaded the Progressive labor movement. Critics claimed white workers were being inhumanely "driven" or reduced to "wage-slaves." But they rarely acknowledged just how deep those connections went. Long before the clock became the cold emblem of Taylor's time studies, it had been a brutal fixture in the hands of plantation overseers, measuring picking rates and—ironically given the later racial narratives on efficiency—calibrating both an individual slave's capacity and their worth.[28] As Caitlin Rosenthal notes, slavery and scientific management had their closest parallel in the task idea. Hailed by Taylor as the "the most prominent single element in modern scientific management," the task idea was one of the principal methods of organizing labor under slavery. Unlike gang work, where slaves labored under constant supervisions for a set amount of time, in task work, slaves were expected to complete a task (usually expressed as a quota) by the end of the day. Workers who achieved their quotas were sometimes given a bonus in the form of money or goods. Those who didn't were punished.[29]

In its incarnation as a tool of scientific management, the task-and-bonus system was elaborated by Taylor's chief lieutenant, Henry Laurence Gantt. Gantt's father had owned more than sixty men, women, and children on his Maryland plantation in the early 1860s.[30] His family claimed a bloodline stretching back to Gilbert de Gantt and the Norman Invasion, but they were more immediately (and more plausibly) related to English and Scottish Protestants who had first settled in Maryland in 1660. After the Gantts' slaves claimed their freedom in January 1863, the family couldn't profitably work its plantation. Creditors eventually foreclosed on the property, and the family moved to Baltimore in 1871.[31] Gantt's father, physically disabled and deaf, was considered unemployable. For income, his mother ran a boarding house before becoming "an invalid," in her son's description, in 1883—a condition that threatened to push her into the ranks of the "idle."[32] She blamed the Civil War and emancipation for destroying her family's social fortunes. Her son, whose high-school principal had served in the Army of Northern Virginia and whose personal hero was Stonewall Jackson, would reflect that family history in his professional preoccupations. His early experiences informed two of the most prominent issues shaping scientific management in the early twentieth century: how to motivate labor—that is, how to make people work in the absence of physical coercion—and a lingering anxiety about the economic and cultural stigma attached to "idleness."

The problem of how to motivate labor crossed the divide separating antebellum plantations and burgeoning postwar Northern foundries and steelworks, where the bulk of scientific management techniques were developed and implemented. But the techniques had to be adapted. In the case of the task-and-bonus system, they most notably needed to address a world filled with machines that didn't always behave as expected. Gantt's efforts to develop and implement the task system under Taylor, and his later break with Taylor on these same grounds, were embedded in the problem of how the selves of free white laborers should be tied to the workings and breakings of the technologies they tended.

The Factory of Failures

From its origins, scientific management was deeply concerned with machines, and not just as models for human efficiency.[33] Taylor had distinguished himself at the Midvale Steel Works as an inventor and technician, creating and refining products that included a railroad car wheel, forging tools, and a steam hammer.[34] He dealt routinely with the breakdowns, malfunctions, and failures that constantly interrupted industrial operations, experiencing them firsthand as a lathe operator in Midvale's machine shop in 1878. Those events increasingly consumed his thoughts and efforts as he rose through the positions of gang boss, foreman, and master mechanic—a position he took on in 1881, the same year he initiated his first time studies. There, a constant stream of department heads darkened his doorway with accounts of debilitating breakdowns and pressed Taylor for urgent repairs. In these years, his concerns over human labor always ran parallel (and often subordinate) to concerns about breakdowns in the lathes, steam engines, and metal presses that populated Midvale. Barely a year before the mania for scientific management would launch him into celebrity, Taylor built his own heroic self-representation through his handling of broken machines at Midvale. He would use the successful redesign of a machine that had previously "battered itself to pieces" as a way of illustrating how his work epitomized "legitimate invention."[35] In the years when his "system" was taking shape, Taylor identified himself, his key lieutenants, and his most important accomplishments with the problems of machine failure.

One of the forgotten elements in this early history of scientific management is that Taylor's earliest efforts at efficiency *produced* breakdowns as a specific kind of material, social, and conceptual problem. The phenomenon wasn't lost on his critics, who saw breakdowns as evidence of Taylorism's immorality. In a scathing critique of Midvale, Samuel Gompers, founder of the American Federation of Labor, characterized Taylor's system as an "industrial slaughterhouse," not just of people but also of machines: "the scheme by which machinery is driven at a pace which destroys it long before its producers guarantee, and where the workmen's health and life-strength are undermined, broken to meet a hastened end."[36] As gang boss and then shop foreman in 1878, Taylor had pushed his machinists to increase production. Workers had rebelled by breaking machines and blaming it on Taylor's system and its punishing pace.[37] Taylor never forgot it. *The Principles of Scientific Management* would mainly invoke breaking machines in one instance: as an act of sabotage against the increased speed that his own system demanded and as a way of masking resistance in the frailties of tools.[38]

But the machinists had also spoken a truth: Taylor's system ran machines, as well as workers, harder, producing breakdowns that weren't easily distinguished from subtle sabotage. Those same breakdowns interfered with Taylor's attempts to reveal the hidden "possibilities of men."[39] They raised two questions that would persist for decades: How should failures figure in scientific management's calculus about the qualities of people? And how could scientific managers make workers identify with, and to be troubled by, the breakdowns of their machines?

Sometime in the summer or fall of 1883, when he left the position of master mechanic to head up the drawing office, Taylor began designing and overseeing construction of a new machine shop north of the Midvale foundry (figure 4.2). He would turn it into a laboratory for experiments on both worker performance and breakdowns, instituting a new wage scheme that would distinguish "high-wage men" and "low-wage men," and launching a nine-year investigation into one of the most widespread forms of machine failure at the time.

In factories across the United States in the late nineteenth and early twentieth centuries, the power for machines came from a long rotating

Figures 4.2
The new Midvale machine shop, completed in 1884.

shaft running the length of the workshop. Known as a "line shaft," it was driven by a steam engine, sometimes enclosed in a separate and dedicated building.[40] The engine drove the line shaft through a series of belts. Individual machines would then take power from the main shaft in the same way (see figure 4.3) through belts that were sometimes thirty feet long, made from strips of leather spliced together to form a continuous loop. Over time, these belts loosened, stretched, slipped, or snapped.[41] Because they bound the factory's power to its production machinery, when the belts failed, entire groups of machines, even the whole factory, could grind to a halt.[42]

Belt failures were one of the most common sources of breakdowns in Progressive Era factories. Harrington Emerson reported an average of twelve belt failures *each day* in the machine shops of the Santa Fe Railway.[43] The caustic lime that hung in the air at paper mills and the carbonic acid that wafted through blacksmith shops ate away at the cowhide belts.[44] At Midvale in the early 1880s, belt failures were the single greatest cause of lost revenue and "by far the largest source of trouble in the shop," according to Taylor. They were so common that a series of "rules" had developed around

Figures 4.3
The shop as seen from the gallery, ca. 1887. Note the "line shaft" running just below the upper row of windows in the top right. Kheel Center for Labor-Management Documentation and Archives, Midvale Company Photographs (1883–1953), Photo ID numbers 5618pb1f4g and 5618PB1F4H.

their repair and maintenance.[45] Taylor's piece-rate system drove the belts even harder, under higher loads, and more continuously, making failures even more frequent and costing time on the machines where they had to be replaced or repaired.

Over the course of nine years, Taylor conducted an ongoing experiment on these breakdowns at Midvale—one of the most exhaustive and comprehensive studies of the subject ever undertaken. A critical motivation behind the experiment was to determine who should be held responsible for the events. In his battle with the machinists between 1878 and 1881, Taylor couldn't always distinguish what had caused a given failure, so he chose to punish individual machine operators. Workers claimed Taylor's incompetence wrecked the machines; Taylor countered by fining them for each breakdown: "Every time you break any part of a machine you will have to pay part of the cost of repairing it or else quit," Taylor declared. "I don't care if the roof falls in and breaks your machine, you will pay all the same."[46]

The fines cost dearly—likely about two dollars, close to a day's pay. The link between machine breakdowns and individual workers now ran indirectly through Taylor's wage system and it helped define working selves. In addition to any fines, workers whose machines broke down would invariably "fail" to earn Taylor's high wage, making them "low-wage men." In the decades that followed, Taylor's disciples would elaborate those connections into a series of accounting mechanisms that tied personal and machine failures, identities and breakdowns, even more closely together.[47]

Birth of a Notation

Henry Gantt elaborated these linkages in more detail and more obsessively than any other efficiency expert of the Progressive Era. Opinions vary on Gantt's character. His close associate and fellow efficiency consultant extraordinaire, Lilian Gilbreth, considered him warm, talkative, and humane. Her daughter thought he was a "cold fish." His voluminous correspondence with his colleague and mentor, Frederick Winslow Taylor, suggests both possibilities. His sharp temper was illustrated on a hunting trip with a group of friends on his family's plantation. When Gantt discovered that their carriage driver, a slave, had eaten their lunch of roasted quail while they hunted, he reportedly erupted in a scene "easily imagined by all who knew him and had heard his explosive comments."[48]

Gantt shared Taylor's lifelong fascination with the operation of machines. He specialized in mathematics as an undergraduate at Johns Hopkins University in the late 1870s. For three years, he taught natural science at his former high school outside Baltimore before pursuing graduate work at Stevens Institute of Technology, finishing in 1885, a year behind Taylor. His coauthored thesis examined the action of different vapors as working fluids in a steam engine. After teaching briefly at the United States Naval Academy, he joined Taylor at the Midvale Steel Works in July 1887 and remained there until 1893, implementing the controversial reforms for which Taylor would become infamous.[49] He eventually became one of Taylor's most trusted lieutenants and a key operative in his mentor's industrial espionage network.[50] After consulting for steel manufacturing plants in Pennsylvania and New Jersey, he rejoined Taylor at Simonds Rolling Machine Company and then at Bethlehem Steel in March 1899. Here, he took up again the question of machine work.

Placed in charge of Bethlehem's gargantuan Machine Shop #2, Gantt's job was to maximize efficiency without the lengthy and detailed time studies that underpinned Taylor's differential pay scheme.[51] He had noticed that perfectly functioning machines were often idle and that workmen always had an excuse ready to hand. To fix this, Gantt proposed changing the way workers and their managers were paid. Acting against Taylor's advice, he abandoned the differential piece-rate scheme, promising a "bonus rate" instead. Workers were guaranteed a daily rate whether they succeeded or failed and an additional bonus if they completed their assigned tasks. If all the machines under the supervision of a manager met their targets, the manager also received a 50 percent bonus, "making it to his interest to give special attention to the men most likely to fall behind."[52] Gantt called the scheme an "education system," and he claimed it taught workers while evaluating and rewarding them.

The new system both masked and extended Taylor's approach of making workers pay for machine failures. Gantt explained, "as it is impossible for the men to earn their bonuses when their machines are out of order, it furnishes *an automatic punishment for breakdowns*, for the man not only loses his bonus on the day the machine breaks down, but on all subsequent days until the machine is running satisfactorily. . . . It is only by the introduction of the 'bonus' plan that we have had furnished the automatic incentive for men to work up to their capacity, and to obtain from the machines the product which they are capable of turning out."[53] Gantt drove the point home, explaining that it was "to the interests of the men to prevent accidents to the machines, and to avoid unnecessary delays," regardless of whether the worker themselves, or other people, were at fault. With the rise of trade unionism and the increasing influence of the International Workers of the World, Gantt believed the system would create a class of skilled workers "entirely distinct from the untrained people we hire" and immune to labor agitation, encouraging workers to believe their economic and moral failure as laborers was tied to the breakdowns of their machines.[54]

In 1908, Gantt began charting the task-and-bonus performance of workers at Brighton Mills, a cotton and fabric manufacturer in Passaic, New Jersey.[55] The mill employed roughly six hundred workers, mostly Poles, Italians, and Slavs, with dwindling numbers of Hungarians, Lithuanians, Greeks, French, and Dutch making up the rest.[56] Cotton arrived at the mill in bales from Egypt and from the American South. It was fed into a series of cleaning machines to produce a long, flat sheet of raw material. It was then carded, combed, drawn, and made into roving. All the work up to this point was performed by the roughly four hundred men who worked at Brighton. The cotton fibers were then spun into thread on bobbins and finally spooled in larger quantities for convenience—the work of nearly two hundred women spoolers. The women worked in large, high rooms "with great belts geared from above, and machines placed in long lanes, where the operatives stand and walk at their work." Humidifying pipes running their length hissed steam from overhead. The air was filled with lint. "The heat, the whir of the machines, the heaviness of the atmosphere, and the lint are at first overpowering to a visitor," one source explained. Many of the women adapted to the conditions. Others lasted only a few days, sometimes mere hours, before leaving.[57]

Gantt had consulted for the mill a few years before, but his work there had been cut short in early 1904 when a cascading sequence of misunderstandings and perceived insults over his bonus system got him fired.[58] In April 1908, with a number of the foremen now gone, Gantt returned and began time studies on the weaving operations. An assistant observed and timed the mill's best weaver, "a Pole named Samtak," and used the results to implement Gantt's bonus scheme for all the weavers, charting the results (figure 4.5).[59]

His graph tracked individual workers, showing on which days they had earned their bonuses. Black indicated success; red indicated "failure." This was no neutral color scheme. At a time when personal failure was intimately tied to notions of debt and insolvency, Gantt's scheme urged workers to get out of the "red" and into the "black." The graphs were also instruments of competition and self-worth. Rather than being locked away in the shop office, they were designed to be displayed in the factory, constantly visible and accessible. In this way, workers would see how their results compared with others and therefore how their bonuses and pay stacked up, spurring pride, competition, and shame.

Two years later, charts in hand, Gantt testified for Louis Brandeis before the Interstate Commerce Commission, explaining how "good men" began coming out of the woodwork because of the charts: "Invariably, in plants where I have been, where they have told me that they did not have any good men in their plants, after we began this system of management, good men began to crop up all around, and in a comparatively short time we had our positions filled with far better men than they ever thought they had there." The charts,

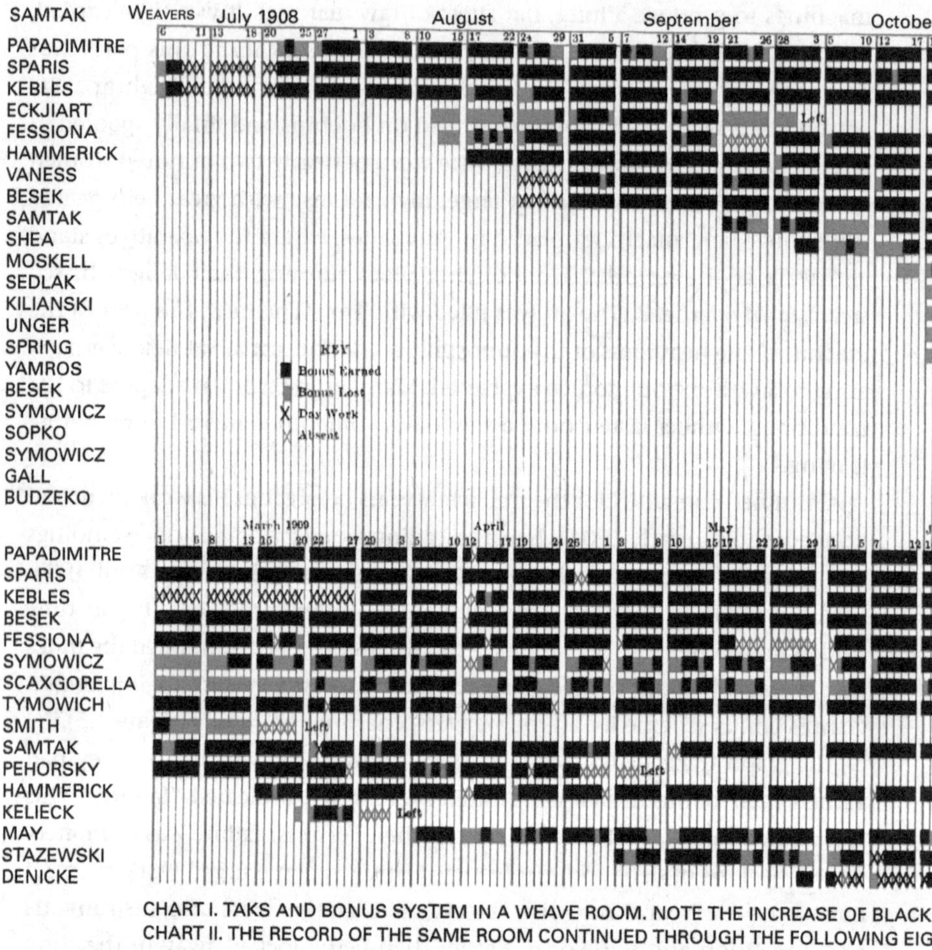

Figure 4.4
Gantt's chart plotting the bonus earnings of weavers at Brighton Mills. The charts tracked individual workers, showing on which days they had earned their bonuses. Black indicated success; red (shown here as gray) indicated "failure." *Source:* H. Gantt, *Work, Wages, and Profits*, 183.

Gantt explained, were revealing the true nature of workers, redeeming them from lapsed lives and forming the basis of personal identities:

> Giving the man a chance, giving him inspiration, makes a man out of him, instead of a machine. He has the same joy and pleasure in his work that the man in the higher position has in his. He has his individual record, as was shown to you on those charts, and he does not like to see a red mark against him. He wants

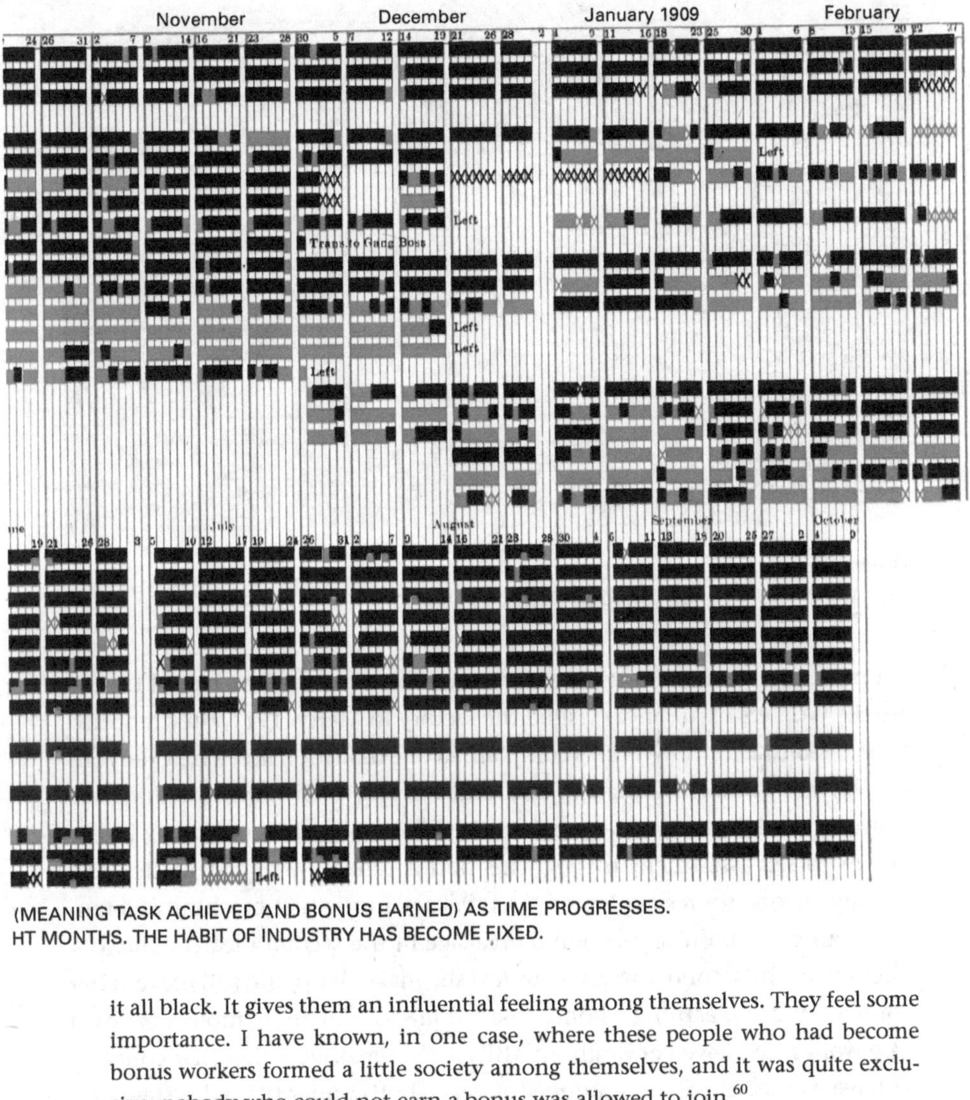

(MEANING TASK ACHIEVED AND BONUS EARNED) AS TIME PROGRESSES.
HT MONTHS. THE HABIT OF INDUSTRY HAS BECOME FIXED.

> it all black. It gives them an influential feeling among themselves. They feel some importance. I have known, in one case, where these people who had become bonus workers formed a little society among themselves, and it was quite exclusive; nobody who could not earn a bonus was allowed to join.[60]

The mill workers, however, spun a very different vision of work under Gantt's scheme. In March 1909, Gantt had extended his bonus wage method to the mill's women spoolers. Within six months, about a third of them had quit. (About a quarter of the male weavers left Brighton Mills over the same period, Gantt alleged, "for some insignificant cause.")[61] Outwardly, Gantt portrayed his system as nurturing and humane. In private letters to Taylor, however, he described it as "survival of the fittest." When Sarah Conboy

Figure 4.5
The Angus Shops of the Canadian Pacific Railway.

(a union organizer and textile worker) and John Golden (president of United Textile Workers of America) visited the factory in March 1911, they reported mass dissatisfaction.[62] Workers complained of exhaustion and a new nervous anxiety about meeting their quotas and earning their bonuses. Alongside slower coworkers and inadequate supplies, Gantt's method of linking breakdowns to self-assessment through his bonus scheme weighed particularly heavily. As one source explained, "A breakage in the first machine of a tandem pair stops both sewers. But a breakage in the second means that work piles up for the second sewer, and unless she makes it up, she will prevent her companion from earning a bonus"—a red mark.[63] An anonymous weaver of fine work interviewed at Brighton Mills listed breakage as her first concern: "I have worked there for a number of years, both under the old system and the new one, and find the work much harder under the new system. If a loom is stopped for any cause, you lose the bonus. . . . It is drive, drive, drive every minute of the day."[64] At the Sayles Bleacheries, a similar scene: "The girls at the machines. . . . stop for 10 minutes in the morning, and clean the machines and clear away the litter around them. The sewing and stooping are monotonous, and the work on bonus here is apt to cause nervousness, because of uncertainty occasioned by frequent breakages in the machines."[65]

Between 1909 and 1911, Gantt built his system through work at companies throughout the American northeast and into Canada, where he consulted at the Angus railroad workshops of the Canadian Pacific Railway (figure 4.5). A two-hundred-acre repair facility, opened in 1902 north of Montreal, the Angus Shops featured their own power plant, hospital, police force, and fire department. If a CPR locomotive broke down anywhere in Canada and could not be repaired on the spot, it was brought here. Seeing the rows of locomotives, sitting literally idle as they waited for repairs, made a profound impression on Gantt.[66] But his techniques were poorly suited to the irregular, unpredictable, and often highly specialized nature of repairs, and his efforts at Angus failed.[67] His experience, though, highlighted two things: that his own system broke down when applied to repair, and that the vice of "idleness," as opposed to the virtue of efficiency, cut across humans and machines and should be the central concern of management consulting and industrial production.

By 1915, five years after the Eastern Rates Case had made scientific management the darling of Progressive Era reformers, Taylor was dead and the movement was splintering. After a House Special Committee investigated the various systems of scientific management in 1911, the Gilbreths, Emerson, and Gantt broke ranks and began stressing the democratic, educational, and "human element" of their systems.[68] Gantt in particular spent the years after Taylor's death working to distance himself from the Efficiency Movement.[69] In 1916, he gathered around himself a splinter faction of the American Society of Mechanical Engineers, rounded out with business leaders, journalists, former clients, and social-reforming clergymen, to form a group he called the "New Machine." A shadowy, amorphous cabal with nebulous technocratic ambitions and headquartered in New York's Singer Building, the group invoked the rapturous language of purification and spiritual redemption in its mission, describing themselves as "a fellowship at deadly enmity with all parasites and pretenders."[70] Their avowed goal was to establish an "aristocracy of the capable" that would fuse business and politics, taking over the leadership of Progressive Era society from know-nothing financiers, plutocratic politicians, and corrupt labor bosses. In their letter to President Woodrow Wilson, soliciting his support (and suggesting how he should run the country), they explained that by taking the "huge and delicate apparatus of industry out of the hands of idlers and wastrels" and delivering it to "those who understand its operations," they could make

America "invincibly strong through the rectification of its business system and the release of the creative faculties of its people."[71] Their "new dynamic order" would transform America and the world through "an all-correlating moral adventure" that viewed politics clearly through the "dissolving and recreative power of this new-born world of work."[72]

As World War I lurched toward a conclusion, Gantt advertised his graphing system as part of this world-civilizational vision. He had elaborated its details during his rise to prominence within the official establishment, conducting efficiency studies of wartime ship production for the Bureau of Ordnance, where he used his charts to organize shell and small-arms manufacturing at the Frankford Arsenal. By the summer of 1917, he felt he had "the whole world by the tail" with a new graphical method that centered idleness rather than efficiency. With the system complete in early 1919, he presented it against the backdrop of world history.[73]

Europe lay in ruins. The Soviet Union stood in chains. If things didn't change, Gantt argued, America was next. The problem, as he saw it, was autocracy. The Efficiency Movement had done nothing to avert the great crises of the world: the rise of the Kaiser, World War I, the Russian Revolution. It had fiddled with profits while civilization burned. In particular, scientific management's great mistake had been to focus on efficiency—on making workers more efficient and on eliminating the inefficient—when the real problem was "the idler."[74] This was true, Gantt asserted, "whether we consider *men* or *machines*."[75] Gantt, who spent much of his free time refighting the "Lost Cause," eulogizing Confederate generals and recreating their battles, leaned ironically on the authority of the Great Emancipator. He quoted Abraham Lincoln from 1847: "The habits of our whole species fall into three great classes—useful labor, useless labor, and idleness. . . . The only remedy for this is to, so far as possible, drive useless labor and idleness out of existence."[76] Socialism, Communism, Bolshevism—they could all be defeated by relying on these principled divisions, Gantt held. What the world needed was a system for eliminating idleness, which meant making its causes specific and visible.[77]

Gantt's personal juxtaposition of sympathy for the Confederacy and praise for Lincoln mirrored the conciliations of D. W. Griffith's recent and controversial film *Birth of a Nation*. Released in 1915, it was the most popular film of its day. Woodrow Wilson (a proud Virginian and Confederate apologist) screened it at the White House. The film painted a sympathetic portrait of Lincoln while celebrating the Lost Cause ideology of the South, where Griffith's father had served as a Confederate general. It built up an architecture of white supremacy, not least through its celebration of the Ku Klux Klan. And its main themes—the fear of Black resurgence and political power, the dangers of miscegenation, the need for unified (white) political leadership to confront the threats to the nation—were an extended exhortation to keep Black people in their place, out of politics, beyond the ordered domestic and political life of white America.

The Great Migration threatened that vision. One of the largest internal migrations in world history, it saw roughly six million Southern Black people move to the North over the twentieth century. About half a million of these had migrated during World War I alone, concentrated in Northern cities and industrial jobs.[78] Writing in "The Negro Migrant" in 1923, Philip H. Brown would praise the recent transformation: "The wonderful metamorphosis now in process, by which cotton-pickers are being transformed into steel workers."[79] But those gains came at great cost, particular during World War I. Recruited into industrial jobs because of industrial labor shortages and the end of European immigration, concentrated in the cities where millions of recent European immigrants had settled and where native-born whites had left positions for the military, Blacks became the target of racial terror as they threatened the entrenched industrial racism of the Progressive Era.[80] During a riot in St. Louis in July 1917, a white mob angered over miscegenation and the use of Black strikebreakers at a packing plant, clubbed, beat, shot, and burned alive 125 Black men, women, and children. Black migration threatened theories of "industrial evolution" that associated Black Americans with agricultural labor and claimed that urbanization led to Black degeneracy.[81] Coupled with scientific management, the accusations of Black people "out of place" became part of national narratives that worked to reimagine, reimpose, and solidify the racial geographies of the period.[82]

In that context, Gantt's civilizational vision expressed a deeply racialized view of idleness. In the English-speaking world, idleness had long been a moral failing. It was close kin to a mortal sin: sloth.[83] By the late eighteenth century, it represented the tenacious preindustrial rhythms and values that threatened the new industrial order and characterized the "uncivilized races."[84] During Gantt's childhood, though, it had taken on a gendered and racialized potency, naming the inefficiency and useless "idle busyness" of women and providing a slander against Black Americans through the image of the idle slave.[85] After the Civil War, it had mutated. Idleness now became something forced on people, particularly white men, through the bankruptcies and financial failures that multiplied in late nineteenth-century America. The exploding ranks of the bankrupt spoke of feeling "useless." Unpaid debts made them idle and unable to fulfill their promises as debtors and patriarchs. In one of the classic inversions of the age, bankrupt whites characterized their own condition as slavery, using "the language of enforced labor to describe a condition of enforced idleness."[86] Gantt's father could have counted himself among them. By the time of World War I, idleness was a characteristic of the Negro and a looming danger for white men, especially those fearing displacement by Black migration.

Rooting his vision in productivist rhetoric and couching it in the language of applied Christianity, Gantt described his remedy for idleness as a "technology" that took the form of a system of management.[87] On Gantt's telling, his charts would record the individual contributions of a cooperative society of equals, where every person, from worker to superintendent, was measured, evaluated, judged, and (crucially) rewarded according to their contribution to the elimination of idleness. His system was personally transformative: It *created* good workers (rather than merely "revealing" them in the crucible of Taylorism). Since these good workers were less likely to leave their jobs than bad ones, the system created a stable, miniature democracy of producers: a self-enclosed productivist society, based in cooperation, and immune to labor disruptions.

Linking the failure of the self with the breakdowns of machines became more central than ever in this scheme. Gantt's new system formed a series of interconnected charts. His earlier graphs of worker performance were now relabeled "Man Charts" (ironically, since women workers had originally furnished their crucial subjects; see figure 4.6). To these, Gantt added a direct analog: the "Machine Chart" (figure 4.7), which followed a parallel logic. For each worker or machine, the workday was divided into five segments. Rather than representing units of time, the divisions represented *amount of work*. Taken together, the segments depicted how much work a "good man" or machine could complete in a workday (say, two hundred pieces of cold-forged steel). A thin horizontal line represented the amount of work *actually done* (say, 150 pieces). The gap between the end of the line and

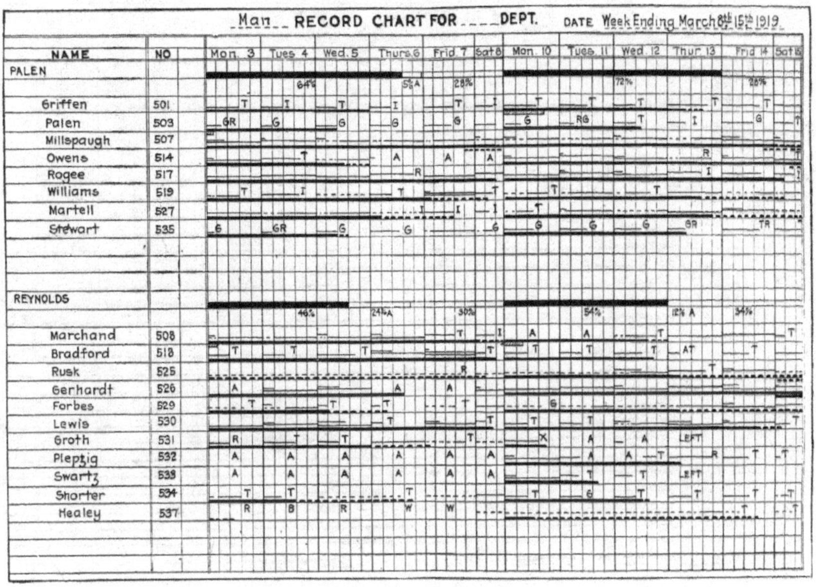

FIG. 8.—KEY FOR MAN RECORD CHART

The daily space represents the amount of work a man should have done in a day, and also the time taken to do the work.
Estimated time for work done.
Time on job for which we have no estimates.
Solid line = cumulative estimated time for work done. Broken line = total time used on work not estimated.
The portion of the daily space through which no line is drawn shows how much the man has fallen behind what he was expected to do. The reasons for his falling behind are indicated by the following symbols:

A Absent — I Lack of instruction — V Holiday
work — M Lack of or defective material — X Reason not clear
erator — T Tool troubles, or lack of tools

Figure 4.6
The "Man Chart." Note the letter codes denoting all causes of "idleness" except for "R" (twelve in total). *Source:* H. L. Gantt *Organizing for Work* (New York: Harcourt, Brace and Howe, 1919).

the end of the day, the unfilled white space, represented "idleness"—a "failure" to perform the proper amount of work. For each week, these daily lines were aggregated into a thicker line, representing potential versus actual work for the six-day work week. Weekly lines were, in turn, aggregated into even thicker lines to give a measure of the manager, represented near the top of the records.[88] Every time that a worker's line filled the day, they were given a bonus. For each day they failed, they were given only their base pay.[89]

In what Gantt saw as his critical innovation, though, each period of idleness—each white space—was labeled with a *cause* so that it could be addressed and eliminated: I (lack of instruction), M (lack of material),

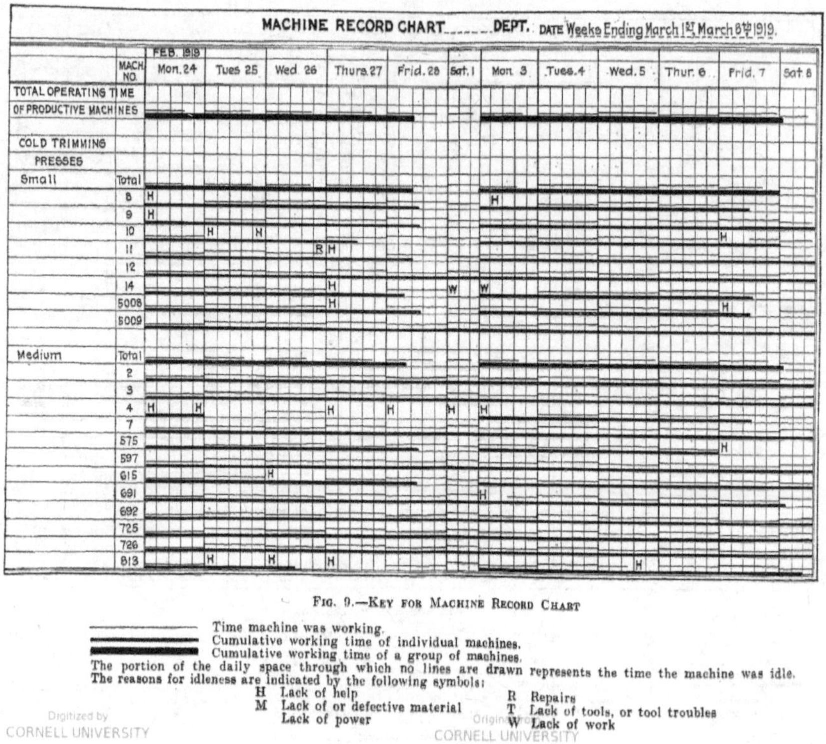

Figure 4.7

The "Machine Chart." Note the letter code denoting "repairs." Also note the single instance of machine failure on Wednesday February 26. *Source:* H. L. Gantt, *Organizing for Work* (New York: Harcourt, Brace and Howe, 1919).

V (holidays), and so on. Looking over the Man Charts for early March 1919, only one letter had no corresponding code on the Man Chart: "R." Its meaning was only decoded by reading it together with the Machine Chart, where it stood for "repairs." The coding signaled exactly how the breakdowns of machines logged on one chart *produced* human failures on the other. Since multiple workers might depend on the operations of a single machine to meet their tasks and earn their bonuses, a single breakdown such as the one in figure 4.7—a failure of the cold trimming press at 11:00 a.m. on February 26—could multiply to produce the dozen breakdown-related human failures in figure 4.6, propagating through the measured output of workers and managers, filling the white spaces that separated achievement from potential, shaking the results of personal output, pay, and worth.

Although hidden in the seemingly objective calculus of the charts, the workplace "democracy" Gantt proposed in fact recommitted itself to America's selective equality at its founding. The industries and enterprises where Gantt consulted—Remington typewriter, Sayles Bleacheries, Brighton Mills, Midvale, the CPR—were dominated by white native-born Americans or Canadians and recent European immigrants. Like so many workplaces, these factories helped transform European workers into white Americans.[90] The Midvale experiment aside, major firms in Philadelphia, for instance, did not employ a single Black worker until the 1930s.[91] Gantt's initiatives may have taken aim at the control of powerful unions but not at their racist practices or their vision of protected white working spaces.[92] Designed to avoid labor disruptions—precisely the conditions under which Black workers often entered the union-controlled shops—his charts buttressed and protected the white worlds of industrial labor. The graphs deployed new "democratic" techniques, five decades after the abolition of slavery, to create one of its afterlives: a system that sought to nurture the mostly male white workforce of the industrial North while reinforcing mechanisms of exclusion that kept Black workers out.[93] That impulse was initially implicit in the graphs themselves. But with the anxieties created by World War I and the Great Migration, under the possibility that Black workers would begin appearing within the closed democracies and have their achievements measured, demonstrated, and supported in the charts, Gantt made the racist scheme explicit.

One of the places where Gantt consulted and where his charts took their final form was the wartime shipyards. The United States Shipping Board estimated that approximately twenty-four thousand Black men worked in these spaces during the war, one-fifth of them in occupations classified as skilled.[94] For the Department of Negro Economics, the shipyards were theaters of Black efficiency. Edward Burwell, a Philadelphia native and captain of a pile-driving crew at the massive Hog Island shipyards, led the team that broke the world's record there in the summer of 1918. Laboring under a torrential downpour, Burwell and his crew drove 220 sixty-five-foot piles in nine hours and five minutes. (The previous record had been 165 piles in nine hours.)[95] Breakdowns figured in Burwell's story only as a minor inconvenience: "Of course we had our little mechanical troubles, but instead of fretting and fuming, the men just glanced at the sign ["when at first you don't succeed, try, try again"] and started in with renewed vigor and the record was smashed." At Sparrows Point, Maryland, Charles Knight, a Black riveter from Baltimore, drove a record 4,875 studs into the hull of a battleship in nine hours. Both the Black and white press praised his efficiency as well as his character—"a highly respectable and industrious citizen."[96] The British Government awarded him $125 for helping the war effort.

Despite the accolades for wartime Black workers, *The New York Times* noted just how stubborn the assumptions of Black inefficiency were: "After [Black workers] had manifested the same evidences of efficiency, they suffered from most invidious discriminations while endeavoring to contribute to the winning of the war."[97] Officials at the War Department ignored these accomplishments and drove the prejudice home: "Clearly the negro has failed to adjust himself to the industrial life and must be returned to his natural occupational environment in the South."[98] Echoing the Southern elites lobbying to reverse Black migration and help Southern agriculture, Major W. Reynolds hoped that deteriorating economic conditions in the North would keep Black workers in the South: "The whole question of course is simply that the darkies seem to prefer to work in the North than in the South. As they find it increasingly difficult to get jobs in the North, things won't seem so attractive there and by contrast, the attractiveness of the South may increase."[99]

Gantt would link these concerns over Black migration to the introduction of his charts. He would have known about the multiracial workforce at the Hog Island Shipyards in Philadelphia. He first trialed his graphs on a mass scale at the Shipping Board and the Emergency Fleet Corporation, where he was a leading consultant. In January 1919, with his graphing technology complete but his book still unpublished, Gantt gave a talk at the De Laval Steel Turbine Company in Trenton, New Jersey. Entitled "Democratic Shop Methods," it announced the final version of his graphs, which he would publish later that year. He began by pinning Germany's defeat on inadequate production, but he quickly turned to what he saw as the problems of American manufacturing. The previous March, he had been at the annual meeting of the Southern Commercial Congress, which had met in his childhood city of Baltimore. There, he claimed, he witnessed a parable of American productivism: "There was a negro lawyer present, who said: 'Gentlemen, this Congress is talking about cotton. But you can't sell cotton unless you raise it, and if you don't keep negroes in the south, you can't raise it. The most important thing for you to do is to see that negroes stay in the south and raise cotton.'" With an almost audible snicker, Gantt added his own cruel ellipsis: "Of course he [the lawyer] was right. The production of cotton was the important thing."[100]

Keeping Blacks in the South meant keeping them under the rule of the Black codes and the quasi-carceral sharecropping system, away from the industrial concerns where Gantt consulted and the employment practices that kept those workplaces white. This fit with Gantt's pastoral vision of Black America and the scenes from his plantation youth. After quoting the anonymous Black lawyer, Gantt explained how America had fallen short of its manufacturing potential for war materiel. The strenuous campaigns for efficiency had not produced results. To illustrate why, Gantt offered another fable:

> I was down south a few days ago, driving along a road in an automobile, when a little ahead of me on a road running parallel were some negroes in a wagon driving a mule. As we approached, the mule heard our automobile and made up his mind that he was not going to allow us to pass him, so he ran away. We soon passed him, however, and then he slowed down to his old gait. Even a mule does not like to be passed from behind; but gives up readily when he realizes he is outclassed. The same law seems to apply to men. A man does not care quite so much if a fellow already ahead of him goes just a little farther ahead, but none of us like to have a man whom we looked upon as our inferior prove to be our superior.[101]

It's unclear whether Gantt (comfortably ensconced in lower Manhattan at this time) was actually in the South "a few days before" his talk. But the Black lawyer from his fable was likely Giles B. Jackson, who had addressed the Southern Commercial Congress in October 1917 in New York and who preached to anyone who would listen that Black people should stay in the South.[102]

Gantt conveniently ignored the counterargument, delivered by William Trent Andrews, another prominent Black lawyer who had spoken precisely on the subject of Negro migration two years earlier. Andrews was born in Sumter, South Carolina, and at the age of twenty-one enlisted at West Point but left because of the poor treatment of Black cadets. He went on to attend Fisk University and then Howard University, where he earned his law degree. In 1886, he passed the South Carolina bar, the only Black lawyer listed. By 1915, according to one report, he held part ownership in nine stores and owned four business buildings and sixty houses, many of them rooming houses, which gave him a sense of both the migrations of Black people and their motivations. William's own trajectory would mirror the migration he analyzed. He would move to New York, where Gantt's New Machine headquarters were housed, and took up a position at the local chapter of the NAACP. His son became a prominent lawyer and correspondent with W. E. B. Du Bois. His daughter-in-law, Regina Anderson—librarian, author, activist—was an important figure in the Harlem Renaissance.[103] But in February 1917, a year before he moved to Baltimore and the site of the Southern Commercial Congress, Andrews spoke at the South Carolina Negro Race Conference on the causes of Black migration.[104]

Andrews laid out the motives behind migration as if presenting a legal brief. The origins of mass migration, for Andrews, lay in the aftermath of slavery. The Black Code had created conditions for Southern Black people worse than the bondage they had just outlived. Reconstruction had failed. The franchise had been granted and then violently taken away. The Black Southerner had "his political rights destroyed, his civil rights menaced, restricted and repressed to a limit as extreme as the federal constitution permits." Every Southern state, Andrews continued, passed statute after statute designed to curtail Black rights, appealing to the white electorate's racial hatred and "future schemes for keeping [the Negro] in his place and crushing his aspiration to measure up to the requirements of American civilization and progress." But Williams soon moved into Gantt's home territory. He outlined precisely how wages, efficiency, and the self were linked in the world of Southern Black people. Suppressed wages meant that Black people couldn't afford ordinary comforts. There was "no stimulus to develop greater efficiency and to improve their condition," which created indifference, shiftlessness, and heedless spending of slim and hard-won resources. Every other class of laborer in the country, he claimed, had been granted an increase in wages except the Black worker in the South. "And still he is charged with being inefficient, lazy, and unreliable. Is it believable that he is ignorant that people of other races receive their true reward for their labor and that his is withheld from him?"[105]

Striking (possibly without knowing it) at the heart of Gantt's motivational scheme, Andrews asserted: "There has never been a time in the history of the world when unwilling workers were classed as efficient; and every man who feels that he is ill paid for his labor is an unwilling worker." He went on to detail the elements that compounded this assault on Black efficiency and the self: Jim Crow railway cars, urban and rural segregation of housing, the segregated textile mills of South Carolina. And he foresaw a worsening where politicians, looking to win votes, championed the poor white working man, exploited racial hatred, and passed laws "to prohibit Negroes from working together in any occupation where white men will want the jobs."[106]

By 1919, Gantt's charts worked to produce precisely the effects that Andrews exposed. His lecture at De Laval positioned the graphs as instruments of continued segregation under new, apparently more democratic systems. His two stories—an anonymous Black lawyer conveniently ventriloquizing for a back-to-the-land movement; and a racial allegory masquerading as a contest between a mule and an automobile—appeared not even a page apart in typescript, but can't be separated. Spoken as Northern cities and industries swelled with Black workers under the Great Migration and returning white soldiers complained about finding their jobs filled, his throwback to antebellum cotton fields, his American technological pastoral, and his charts were about the definition of a white technological self, purged of idleness and underpinning a renewed American civilization. The mule in his parable was a proxy for the real object of ridicule: pre-industrial rural Black people thinking they could compete with modern industrialized whites. The casual inclusion of race in a sermon on industrial production and the incoherent lessons Gantt drew for scientific management only underlined the real purpose of the episode: to cast the graphs as the central instruments in a system of both Black containment and white uplift that would eliminate the scourge of idleness that haunted Gantt's own family.

Conclusion

Under the pressures of the Great Migration and World War I, Gantt transformed his charts into an instrument for linking breaking machines and failing white men so that both could be repaired. His charts were an attempt to make white workers feel the breakdowns of their machines, but not too sharply, shielding them from their full effects through education and uplift. Outside the walls of the factories and workshops where Gantt consulted, those supports were invisible. White workers ostensibly flourished on their own merits. Black workers apparently failed because of the incongruency between their inner worlds and industrial machines.[107] For both groups, though, the charts were the product of both a personal and a cultural imagination—resourceful, active, anxious—a dystopia in which others were forced to live.[108] White American and immigrant workers had little choice but to live in the Progressive Era vision of productivity that Gantt's charts embodied. Black workers also inhabited the world created by these graphs and their logics, left out of efficiency metrics that judged them unworthy of inclusion. If the graphs measured potential, then these were people who failed even to register, whose inner lives and outer performances could not be linked to the properties of machines.

Far from an unintended side effect of scientific management theories, racism was part of the innovative, systemic, and forward-looking visions of its proponents. Appearing alongside eugenicist arguments that would have targeted Gantt's own parents, and by extension Gantt himself, the charts helped to produce the "unequal differentiation of human value" that drove the racial capitalism of the Progressive Era.[109] And breakdowns were at their core. The irony of scientific management techniques, so many of which were drawn from the management practices of slavery and extended by scions of dispossessed plantation owners, was that they were now turned toward a putative advancement of white workers that encouraged them to make breakdowns a problem of the self. For the Black Americans left out of these schemes, this might seem like an escape from the quantitative and captivating logics of capitalism. But that exclusion also shaped individual futures and societal visions of the relationship between Blackness and modern technologies. Gantt charts, together with the contexts in which they originally operated, formed a mechanism through which technological virtues were identified with ostensible racial virtues and through which ideas about technology were created of a piece with ideas about race and the self.[110] Gantt's charts of machinists and bleachery workers, running along the broken, gendered, and colored lines of American labor and society, registered who counted as a candidate for failure and who didn't deserve the chance. It also spoke to the future: to a growing collection of increasingly sophisticated techniques that sought to make others internalize the breakdowns that surrounded them.

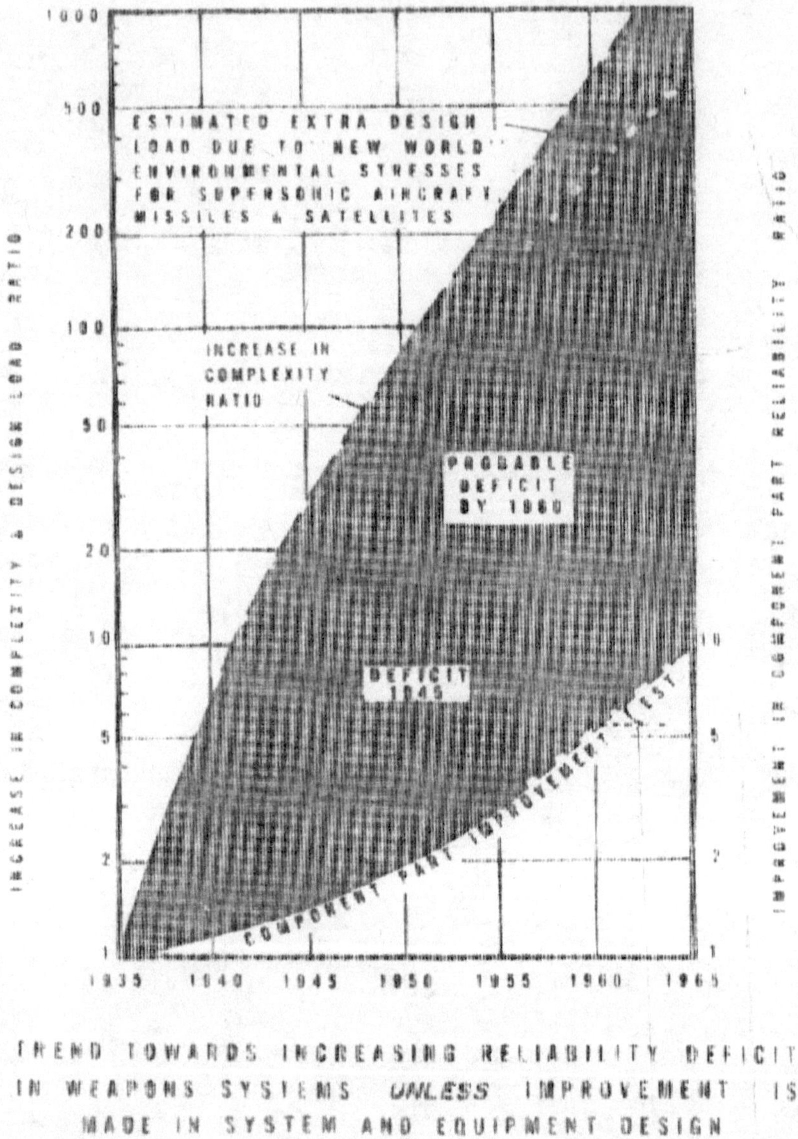

Figure 5.1

The reliability deficit. The shaded area represents the reliability "gap" (sometimes called the "reliability deficit"). Note the uppermost text block, signaling special threats to supersonic aircraft, missiles, and satellites. *Source:* Reproduced from Chester Soucy, "The 'Project MATURE' Concept of the Design and Production Requirements for Reliable and Maintainable Military Electronics Equipment," figure 2; National Archives of Canada RG 24, accession 1983-84/049, vol. 1664, file 1950-123-7, part 2. Courtesy of the National Archives of Canada.

5 Minds and Machines: Cold War Electronics and the Crises of Complexity

Something failed: a man, a machine. . . . It doesn't matter now.
—*Failsafe*, 1964

October 1958. A conference room in Ottawa, Canada. It was from a projection screen, in a non-descript conference room, that the secret graph first escaped the eddies and currents of classified Western military documents and began circulating publicly. This was an age of self-proclaimed crises for the West. Earlier in the decade, the Battle of Điện Biên Phủ had ended colonial rule in French Indochina. The Algerian War of Independence, supported by an exiled Frantz Fanon, had taken down the French Fourth Republic. Egyptian nationalism in the Suez had exposed the naked duplicity of British and French involvement in the region, dividing the Western powers. And the Soviet launch of a R7 ICBM in October 1957, carrying the Sputnik satellite, had been met with launchpad explosions and catastrophic failures as the only Western reply. This graph raised the stakes still further. Even without knowing its details, you could tell it foretold another crisis.

In stark tones, the image captured the potential catastrophe of electronic breakdowns in the early Cold War. It illustrated how the skyrocketing complexity of military electronics had outstripped the reliability of the components making them up. And it projected that growing gap into the future. Between the two curves, between necessity and reality, yawned a chasm—a "reliability deficit"—that threatened the technological foundations of Western military strategy and North American continental defense. Additional reports confirmed its fears. One devastating assessment predicted that Western intercontinental ballistic missiles (ICBMs), the very technologies keeping the Cold War "cold," would "play little or no part in a possible war *tomorrow*, and as we expanded the use of our complex equipment across the boundaries of the world, we would be faced with a dismaying and disastrous number of reliability failures."[1] By the early 1960s, films such as *Failsafe* and *Dr. Strangelove* were darkly and publicly mocking the hubris of Western defense systems, peeling away the reassuring veneer that they would work at all when called upon. This conflict between cold war fantasies of complex, automated, and infallible machines, on one side, and the reality that Western defense and weapons systems were approaching collapse, on the other, generated what the conference graph left unnamed but what contemporary observers gave an ominous title: the reliability crisis.

Histories of Cold War technology have mostly overlooked the reliability crisis.[2] Instead, they have portrayed 1950s electronics as the original "disruptive technologies" of the twentieth century—the protagonists of a material revolution in faster, more reliable, solid-state electronics, inextricably tied to the origins of Silicon Valley and the tech industry's model of infinite growth and constant, radical innovation.[3] But as the conference graph labored to point out, novelty and material crisis were deeply entwined in the 1950s and nowhere more than in the electronic breakdowns of the time. Even before the graph appeared, a patent drawing had gestured toward that enmeshed history. Filed by a coinventor of the transistor and a central figure in the story of Silicon Valley, William Shockley, it depicted the electronic feedback circuits of a self-guided bomb (figure 5.2). In doing so, it positioned Shockley and the early history of Silicon Valley as part of the emerging *problem* of Cold War electronics: an early avatar for the enchantment with automation that would soon fuel both the reliability crisis and the birth of the Valley itself.[4] Rather than two isolated and diverging documents, conference graph and patent drawing were two sides of the same sheaf of paper, illustrations in a single intertwined history of complexity, automation, and electronic breakdowns in the Cold War.

This chapter explores this history by returning the reliability crisis to the central position it occupied in the late 1950s. I argue that this crisis, shaped by the felt vulnerabilities of Western nations operating within a "hostile" globe, once again transformed breakdowns into sites for crafting specific visions of the self, this time in relation to the proper citizens of the Cold War West. It did this in two ways. By framing breakdowns as a problem of complexity, the reliability crisis linked the work of North American technologists with social scientific concerns about the challenges facing

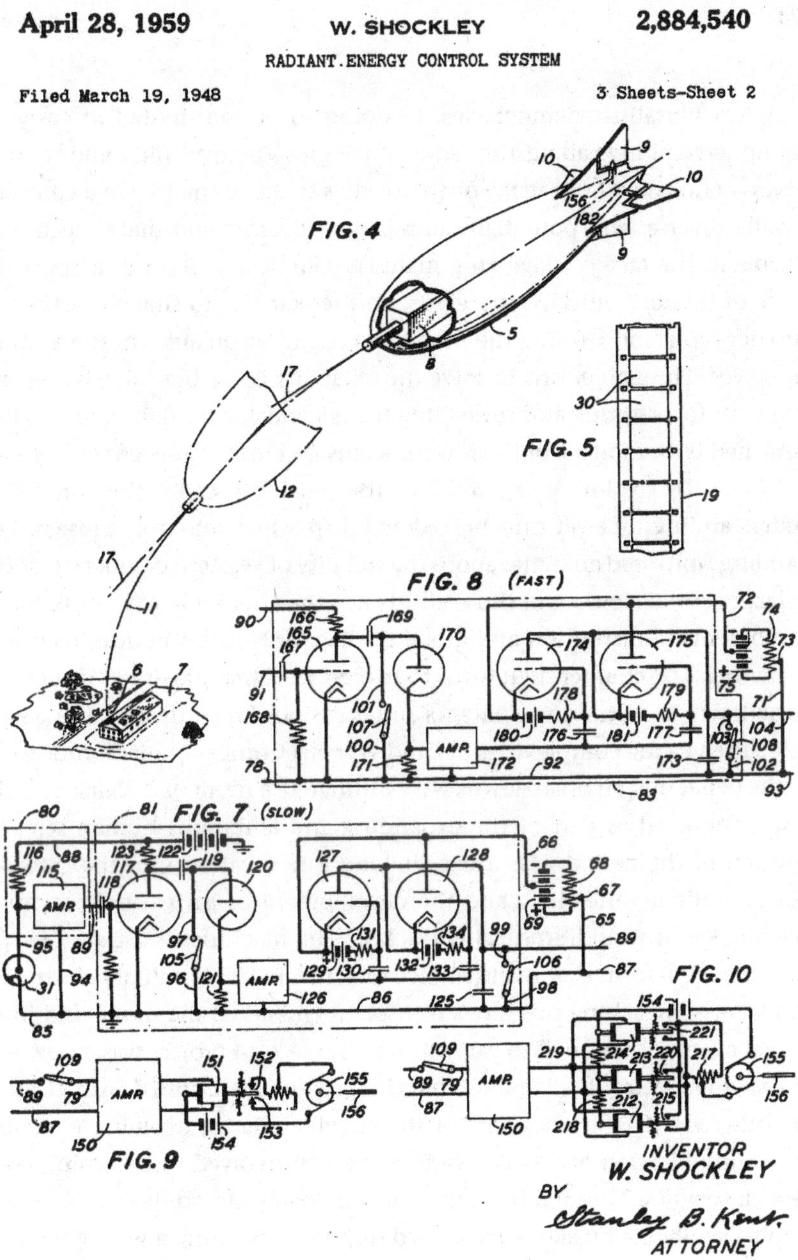

Figure 5.2

Part of William Shockley's patent for the electronic servo-mechanism in a self-guided bomb. The patent was originally filed in 1948 while Shockley was working for Bell Laboratories, where he coinvented the transistor. *Source:* William Shockley, 1959. Radiant Energy Control System. US Patent 2,884,540, filed March 19, 1948, and issued April 28, 1959.

complex, pluralistic democracies. In doing so, it contributed to converging understandings about the kinds of people—the reasoning and rational selves—who could both make open societies stable in the face of a complex, socially diverse, and potentially unreliable citizenry, and make electronics reliable in the face of staggering material complexity. Second, it spurred a range of military officials, strategists, and researchers to shape breakdowns into occasions for defining the moral and cognitive qualities of these thinking selves. Through efforts to solve the reliability crisis, breakdowns became moments for asserting and cultivating the dependable thinking subjects that furnished both model Cold War technicians and model Western citizens.

I begin by exploring the roots of the reliability crisis, the competing understandings of electronic breakdown it spawned, and how those understandings mirrored anxieties about the stability of Western democratic social order. A central element of those efforts was the attempt to cast breakdowns as moments of disruption and moral display for North American technical personnel. The chapter then turns to two projects that illustrate the place of breakdowns as disturbing elements and occasions for cultivating the people best suited to the complexities of Cold War electronics—projects motivated by the belief that all breakdowns were ultimately a result of *human error*. The first, conducted as part of the expanding and ambitious human sciences research of the period, was a decade-long interrogation into the category of electronic maintenance and the development of increasingly complex machines that would, ironically, teach it. The machines illustrate the specific historical form that maintenance took in late 1950s military electronics and how the teaching machines it inspired embodied the moral and intellectual agents envisioned to carry it out. The second project was a new way of drawing electronic diagrams that, its proponents claimed, would guide the otherwise fallible thoughts of the circuit designer, avoiding error and miscommunication across the various groups involved in designing complex electronics. Through both these technologies—teaching machines and graphic symbols—breakdowns formed sites for cultivating a way of being in relation to the catastrophically fallible technologies of the Cold War.

The New World Order

The reliability crisis crystallized a series of ongoing debates in the 1950s over exactly *why* electronic equipment broke down, debates that reverberated with analogies for the robustness of pluralist, democratic societies. Early in the decade, the most influential view came from the US ballistic missile program at Redstone Arsenal. Its main proponent, Robert Lusser, was a former German aviator, aircraft engineer for Messerschmitt and Heinkel, fierce rival of Werner von Braun, and codesigner of the V1 under the Nazis. Lusser was brought to the United States as part of Operation Paperclip, the secret US intelligence program that employed captured German scientists on advanced weapons and engineering projects after World War II, including the US ballistic missile program at Redstone. From his position as head of the Reliability Branch there, Lusser complained about a wave of automation "sweeping the world." He warned that the electronic equipment underpinning automation was "notoriously unreliable" and posed a threat to the operational readiness of many of the West's major weapons systems.[5]

Lusser's answer to the problem of technological failure took the form of a mathematical formula that came to be known as Lusser's law. It held that the reliability of an electronic assembly—a radar set, for example—was defined by the mathematical product of the reliabilities of its constituent parts.[6] Breakdowns were therefore a problem of flawed individual constituents within a larger whole. Lusser described the condition in the language of human frailty: the "weakness" of individual parts, the "diseases" they suffered from, the high "infant mortality" of new components. Each individual part—each "atomic" constituent—was a potential breakdown waiting to happen. As equipment grew more complex, these failure points multiplied, their weaknesses proliferated, and overall reliability plummeted. Complexity, therefore, was "the most prominent cause of unreliability." Prolonging the life of an assembly meant not only making electronics simpler but also making these individual components "stronger."[7]

Lusser's pathology-inflected understanding of breakdown was also a statement about the *nature* of complexity, specifically its relationship to uniformity and difference. In a 1957 report, he depicted two assemblies: one "homogeneous," composed entirely of identical parts, and the other "heterogeneous," with no two components the same (figure 5.3). Assuming there were sixty-three components in each assembly and that any single component had ten possible ways of failing, Lusser explained that the heterogeneous assembly would suffer from ten times as many failure modes as the homogeneous equipment. Envisioning the assembly as a social microcosm, a move made more troubling given his wartime activities under

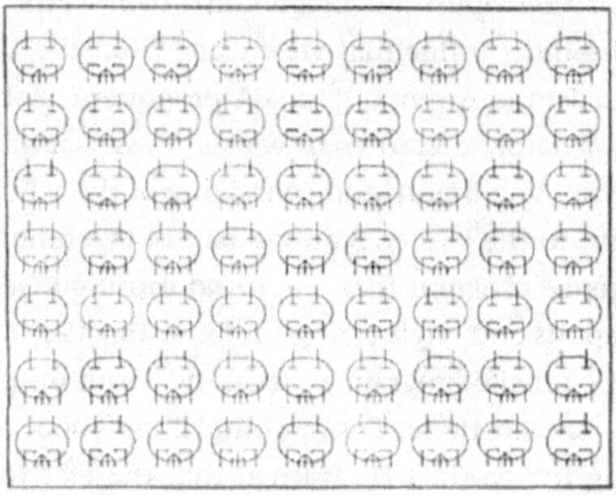

Fig. 7 Homogeneous Assembly
63 Equal Components
10 Modes of Failure

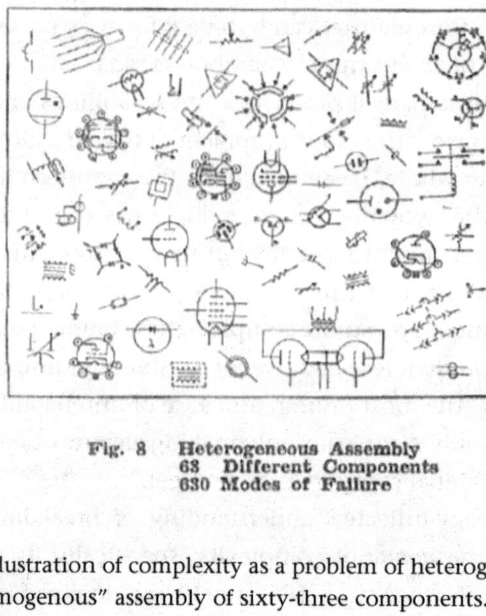

Fig. 8 Heterogeneous Assembly
63 Different Components
630 Modes of Failure

Figure 5.3
Robert Lusser's illustration of complexity as a problem of heterogeneity. The top figure shows a "homogenous" assembly of sixty-three components. The bottom figure shows a "heterogeneous" assembly of the same number. Each component is assumed to have ten modes of failure. Lusser noted that, under these conditions, the heterogenous assembly would have ten times as many modes of failure. *Source:* Robert Lusser, "Unreliability of Electronics: Cause and Cure," Research and Development Division, Ordnance Missile Laboratories, Redstone Arsenal, November 1957.

the Third Reich, the cantankerous Lusser emphasized *in his own italics* the approach that heterogeneity demanded: "If we really want to increase reliability . . . we must consider all of the types of components as being *potential criminals*." (Lusser's V1 work at Peenemunde had used prison labor from the SS concentration camp at Buchenwald.) The way to interrogate these elements, forcing them to reveal their weakness and ultimately making them "stronger," was "by individual tests to failure."[8]

Starting in the late 1950s, critics of this atomic approach to breakdowns began arguing that the focus on simplicity and robust individual electronic parts was misguided. Complexity, they felt, was a necessary and defining characteristic of modern technologies. An assembly made entirely of identical electronic parts, after all, performed no useful electronic function at all. The question was how to make complex electronics function stably when their constituent elements were both varied *and* fallible. As the author of the Ottawa conference graph would put it, "Our major military problem is not, as it is sometimes urged, that of making equipment simpler—desirable though this obviously is—but, more realistically . . . of making necessarily complex equipment more reliable."[9]

The graph itself was a weapon in this last campaign. Its author was an engineer and member of the Canadian Air Material Command named Chester Soucy. An Air Force signals and radar officer during World War II—a position that made Robert Lusser his direct and immediate wartime enemy—Soucy was also chair of the local parish board of his Unitarian Church and known for his controversial homilies.[10] His graph announced nothing less than a schism.

Soucy argued that breakdowns were largely the product of a lapsed morality. Among reliability experts in the late 1950s, this was uncontroversial. A consensus had developed around the belief that technical personnel were not sufficiently *troubled* by the specter of breakdowns. Lusser himself had blamed malfunctioning avionics on the moral failings of designers who, he felt, failed to take a sufficiently "life-and-death" approach to their work.[11] His allies at Redstone had used that accusation to excoriate system designers and their "supercolossal gadgetry."[12] Carrying Lusser's language of criminality from electronic assemblies to the irresponsible people designing them and the breakdowns they produced, one Redstone official explained that there had "been more crimes committed under the concept of system engineering than any other crusade. . . . The system is only as good as its weakest component."[13] Soucy instead turned this charge of immorality against Lusser's own philosophy, arguing that only a broader vision of breakdowns could responsibly confront their scope and danger. The language of Soucy's faction was proselytizing. It openly preached a new "gospel of reliability," openly called for the "indoctrination" of officials and technical staff, and energetically proposed a new morality, focused on making breakdowns a troubling element in the inner worlds of the technical personnel of the armed services of the West.[14]

But Soucy's program followed this Cold War emphasis on morality, fealty, and ideological purity—failings of the soul—with concerns over the human mind. In one of the period's most sweeping claims about electronic breakdowns, he wrote: "The common factor in all causes of unreliability is *human error*, whether it occurs in the original design, fabrication, inspection, maintenance, or operating use."[15] Soucy's illustrations drove the point home, locating the source of breakdowns outside the individual electronic parts that Lusser had criminalized and encasing them graphically within the larger human environments that produced them (see figure 5.4). By framing complexity as both threatening and necessary, reliability as a moral crusade, and breakdowns as a problem solved by human thinking, Soucy's chart wove electronic failures into the broader challenges facing Western societies and the salvational role of the thinking self.

Since the end of World War II, the Cold War social sciences had also cast complexity as a central and pressing challenge for Western democracies—a defining feature, a marker of plurality and diversity, and an existential threat to social stability and cultural unity.[16] Drawing on the technical sciences for their metaphors and models, social scientists interwove solutions to the problems of advanced technology with solutions to the problems of social order.[17] Their efforts focused in particular on defining what kinds of individual constituent elements—in this case, what kinds of selves—might ideally withstand the stresses and challenges that the complexity of modern societies generated. As Jamie Cohen-Cole has noted, the "open mind"—tolerant, creative, adaptable, emotionally stable—provided one powerful and attractive solution.[18] But in the all-important sphere of decision-making, where complexity was the parent of error, solutions circled consistently around the virtues of reason and rationality, which would cultivate the kinds of people capable of cutting through dizzyingly complex information and contexts using sound judgment, previous experience, rigorous logic, formal rules, and streamlined algorithms to avoid potentially catastrophic errors.[19]

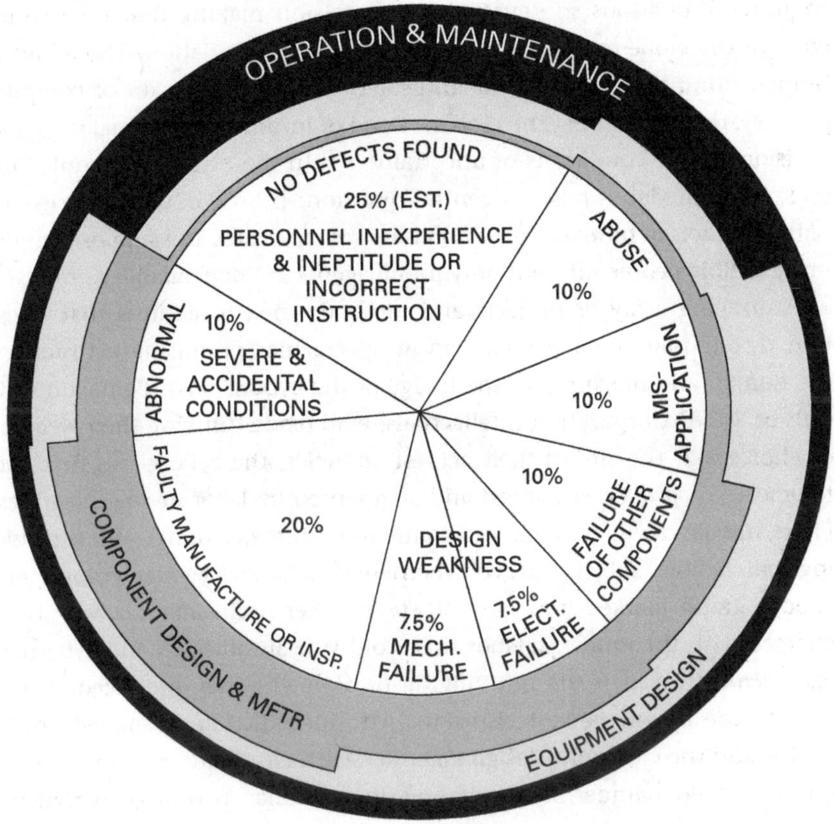

Figure 5.4
The human causes of failure. The chart shows electronic failures in military equipment (center) encased in the human practices of operation, design, and production that control them. Note the emphasis on design weakness, misapplication, abuse, and "no defects found" due to maintenance inexperience, ineptitude, or incorrect instruction. *Source:* Chester Soucy, "A Broad Survey of the Military Electronic Equipment Reliability Problem," *Proceedings, 1956 Electronic Components Symposium* (New York: Engineering Publishers, 1956), 8–23. fig. 5. © (1956).

This emphasis on right thinking and the dangers of error flourished in the Canadian military circles where Soucy moved—a collection of bureaucrats, strategists, and defense intellectuals who introduced the concept of technologically "assured destruction" in 1951 and elaborated the doctrine of first- and second-strike nuclear forces.[20] But they represented a much broader anxiety among Western officials. As Joy Rohde notes, Cold War intellectuals were "profoundly anxious that humans suffered from innate cognitive limitations so devastating to decision making that they could consign the globe to endless wars, even atomic annihilation. The unaided human mind was incapable of understanding the intricacies of complex social systems, absorbing the modern flood of information, or making good decisions under conditions of uncertainty."[21] In the case of electronic failures, those anxieties raised a central question: If breakdowns were essentially products of human error, what kinds of structures and supports might make fallible people into not only moral agents but dependable thinkers?[22]

A dizzying array of projects and methods tried to address that question throughout the 1950s, the most spectacular promising a fusion of the human and machinic in the image of the cyborg.[23] For Pentagon officials or RAND Corporation intellectuals, who made thinking their weapon of choice and the site of their acutest anxieties, the cyborg was first and foremost an *operator*, enhanced and augmented by labor-saving machines. Pilots, missile crews, gunners, and command center personnel—these thinking, but fallible, subjects drove cybernetics' most spectacular visions. But Soucy's graph marked how, just off-stage, other problematic selves stood charged with the immense labor that Cold War automation and cybernetics *generated* through the multiplying breakdowns they produced. Those others—the maintenance technicians who identified and repaired breakdowns and the electronic designers tasked with eliminating them—are the focus of the remainder of the chapter. In particular, two projects that lay behind Soucy's graph—one to automate the teaching of maintenance and the other to failure-proof the act of electronic design—illustrate how breakdowns figured as Cold War sites for cultivating precisely the kinds of selves who would care about and combat them.

Maintaining Machines

As much as the reliability crisis sparked questions about what caused breakdowns, it also spurred tortured questions about maintenance and repair. In tackling those questions, Soucy and his allies drew heavily on studies of electronic maintenance conducted earlier in the decade and particularly on one influential project for the US Navy conducted in 1951. Following years of reports that naval electronic equipment was massively unreliable, the psychology department of the University of Southern California (USC) sent two-man research teams onto thirty US Navy vessels, most likely operating out of the Long Beach Naval Yard, during training operations.[24] While technicians literally looked the other way, the researchers "injected" faults into live and fully operational electronic equipment (replacing perfectly good resistors, for instance, with faulty ones—Lusser's "potential criminals"). They then observed "every single action" that technicians performed trying to find and fix the faults. Since introducing malfunctions and breakdowns into active military electronics was a questionable research strategy, the USC team soon decided "to put the bird in the cage and carry it to the laboratory."[25] A year later, the Office of Naval Research contracted USC's psychology department to establish an Electronics Personnel Research Group focused on human aspects of maintenance. It launched one of the largest and most widespread interrogations of maintenance in history and would inform almost all North American military thinking on the subject for the next decade, including Soucy's.[26]

The USC experiment challenges two tendencies in recent histories of maintenance. One is to see maintenance work as largely hidden, occupying what David Edgerton describes as "a twilight world, hardly visible in the accounts societies make of themselves."[27] As a growing literature suggests, however, many societies, including in the Global South, make acts of maintenance and repair central to their self-understandings.[28] Specific historical periods have also seen maintenance and repair become central and even defining cultural features. Within Western defense establishments of the early Cold War, the reliability crisis made maintenance *hypervisible*, not only as a "working object" of the burgeoning postwar human sciences but also as a central feature in the accounts that cold warriors gave of their advanced technological society and its prospects for survival.

The USC investigations also challenge the tendency among scholars to treat maintenance as a transhistorical phenomenon. "The dramas of maintenance and upkeep played themselves out in ancient history," Lee Vinsel and Andrew Russell have explained. In arguing for its universality, they continue: "Maintenance was fundamental to all societies, including the most basic human settlements, like the stick shelters of the San people of southern Africa."[29] In its ambiguous use of the past tense to describe the still-existent San, that formulation draws on long-standing tropes that use non-Western, and specifically African, peoples to stand in for a timeless humanity.[30] With even "the most basic" human societies engaging in its practices, this vision of maintenance conflates an analytical category with a universal and ahistorical activity: "the distinctive forms of work that go into keeping things the same."[31] But precisely because it presumes "sameness" as the goal, assembling practices as diverse as washing and mending to overhauling and rebuilding into a single global activity, that view of maintenance treats cultural forms with radically different histories, social relations, and meanings as the same kind of historical object, changing only in scale, scope, and form, as human societies move from "basic settlements" to advanced industrial conglomerations.

As a way of challenging the innovation-based histories of Cold War electronics, the focus on maintenance reveals much. But the understanding of maintenance outlined above doesn't just replace one kind of unhelpful progression with another. It is, arguably, a *product* of exactly the kinds of Western global-scale ambitions and concerns that the USC experiments participated in. In other words, this way of framing maintenance avoids historicizing our own understanding of it, part of which is a product of the form it took during the Cold War.[32] When did maintaining an object become synonymous with repairing, mending, caring for it, or keeping it "the same"? When and where did it take on, or take on once again, its combined material and moral dimensions? And how did historical actors themselves understand these forms of work, the technologies they touched, and their wider relations in their own time? As an interrogation into the nature of maintenance itself, the USC experiment reminds us to not take the historical stability of that category, and our contemporary assumptions about it, for granted.

Maintenance researchers within Cold War North American militaries, for example, generally saw electronic breakdowns as a radical rupture from anything that afflicted machines like the eighteenth-century guillotine or nineteenth-century cotton gins or the steel presses of the Progressive Era. To begin with, the *way* electronics failed, they argued, was new in the history of technologies. Mechanical contrivances tended to wear down gradually. Electronics, on the other hand, seemed to fail suddenly and randomly.[33] Those failures often thwarted the naked human senses. The careful eye or ear of the Victorian locomotive engineer might sense the telltale signs of a weakened link rod or a loosened bolt. Electronic devices, on the other hand, hid their intentions until the last minute and generally maintained the same outward appearance whether they were operational or not. They followed linear laws that befuddled the logarithmic hearing and sight of humans, who had to rely on additional test equipment, itself fallible, to detect problems.[34] And then there was the complication of complexity. Officials in the 1950s reported that defense systems had become so complex that existing technicians could not even find the sources of equipment breakdowns, much less repair them.[35]

The USC investigations off Long Beach aimed to describe and codify the behavior of successful technicians so that it could be quickly and efficiently taught. The hope was that this would help solve a growing manpower shortage.[36] Those expectations were quickly shattered, though. Breakdowns, the USC researchers found, did not launch an obvious and universal "protocol"—a sequence of logical processes and actions leading to repair. There instead seemed to be no obvious logic to technicians' work. Troubleshooting followed no fixed sequence. Technicians rarely used manuals at all, and when they did, it was not systematic. They tended not to use circuit relations or mathematical formulas. In fact, they almost never did any calculations at all. Their supervisors saw technical troubleshooting as a matter of individual skills, a task that only highly trained and experienced personnel could effectively perform. Almost every maintenance "performance" included errors in equipment use, which usually had little influence on a solution. And perhaps most troubling, inexperienced technicians seemed to *generate* electronic malfunctions as a matter of course.[37]

Despite those findings, USC researchers set out to develop teachable troubleshooting protocols. At their laboratory on the USC campus, and later at the Navy Electronics Training School on Treasure Island, in San Francisco Bay, they developed teaching technologies to help with this instruction. These eventually took the form of "teaching machines."[38] The teaching machines turned the event of the breakdown into a mystery, positioned it at the center of a web of selectively revealed clues, and based the solution on the successful troubleshooting strategies of a fabricated "ghost expert," whose choices were built into the functioning of devices. The structured approach of the ghost expert broke down the problem-solving into phases—symptom recognition, hypothesis formation, and hypothesis testing. Selectively revealing information and guiding trainees through correct and incorrect choices, the teaching machines tested a student's ability to solve the puzzle of the breakdown, working to align their thinking and choices with those of the "ghost expert," whose solutions represented the only "correct" possibility.[39]

The maintenance machines were part of a 1950s explosion of interest in devices that promised to automate teaching, provide immediate feedback, and foster individualized learning as a response to Soviet dominance in science and technology.[40] The US military had been early and enthusiastic patrons of these projects, funding Sidney Pressey's development of an "auto-tutor" for the navy in the early 1950s and supporting a five-year project that led to B. F. Skinner's "science of learning and art of teaching" in 1954.[41] But the machines also positioned their students within a longer tradition of techniques for cultivating democratic citizens.[42] Maintenance studies regularly drew on the educational philosophy of John Dewey, the pragmatist philosopher, educational reformer, and public intellectual, who had also proposed problem-solving as a way of forming thoughtful democratic citizens capable of reasoning and adapting to a complex world.[43] And like Dewey's vision, the teaching machine projects (including the ones designed for maintenance) made crucial assumptions about their learners. As Tiago Saraiva and Cornel West have observed, although Dewey cast his problem solvers as universal, they had "the flavor of small-scale, homogeneous communities."[44] The postwar teaching machines may have embodied a different conception of the democratic citizen than Dewey's, but creating a machine whose subjects were charged with the nation's most advanced technologies made its own assumption about who these learners were.

In framing the problem solver as a race-less and gender-less abstraction, the maintenance machines inevitably cultivated a gendered and racialized vision of both the "ghost expert" and the technicians he would teach.[45] The assumptions of the male technician in particular belied the contemporary history of advanced electronics. As Jennifer Light noted in a now-classic study, the women programmers of the wartime ENIAC computer had been charged with not only programming the device but also analyzing and fixing its frequent breakdowns, sometimes tracking down burned-out vacuum tubes in the machine's room-sized chassis.[46] The "restructuring" of the early Cold War computing industry forced women out of analogous roles in Britain, with devastating consequences for its electronics sector.[47] Drawing on this purged workforce for their "ghost experts," and using them to train new recruits, the maintenance machines obscured this history, redefined who counted as a good troubleshooter, and drew increasingly sharp lines between the feminized work of electronic production and the contemporary, male-dominated work of problem-based electronic repair.[48]

Images such as figure 5.5 illustrated this shift and its contradictions. Both Soucy and Lusser, for all their differences, had presented the production of

Figure 5.5
An unidentified engineer wires an IBM computer in 1959. The photo illustrates increasingly sharp lines between the increasingly feminized work of electronic production and the ostensibly male-dominated work of electronic faultfinding and repair. *Source:* Getty Images ARC0005735_078. *Credit:* Berenice Abbott.

complex electronics as a comparatively easy task—a factor that contributed to the chaos of automation. Tracking the subtle indicators of breakdowns through the tangle of all-black wire in a 1950s SAGE console, however, was something only the expert, increasingly male, logical problem solvers being created by the maintenance machines could hope to accomplish.[49]

Drawing on those perspectives, Soucy himself positioned teaching machines as a solution to both the practical and moral dimensions of breakdowns. First, the machines provided an instrument to attack Lusser's simplistic calculus of reliability. Because they allowed testing in controlled settings, Soucy believed the machines could quantify the technician's tendency to commit errors, allowing officials to construct a "maintainability index" and a new calculus of reliability to replace Lusser's law.[50] Soucy would also come to see the teaching machines as an instrument for looking into the souls of technicians, determining how strongly they internalized these breakdowns as problems along with their trustworthiness in resolving them.

Since at least the time of the USC studies, Cold War maintenance had been steeped in suspicion. Echoing Frederick Winslow Taylor's accusations against the machine shop operators at Midvale, military officials in the 1950s often saw a blurred line between breakdown and sabotage. The authors of the controversial book *The Crisis We Face*—a tirade against the effects of unbridled automation—expressed a widespread suspicion, for instance, when they imagined the following scenario: A Soviet agent, installed at a key missile production facility and armed with counterfeit parts that were designed to fail, could insert them into ICBMs in such a way that the missiles would pass tests on the ground but break down during flight. Remember, they implored their readers, "hundreds of maintenance men working on hundreds of thousands of pieces in each missile—one of them can sabotage a squadron."[51]

The Cold War social sciences trafficked heavily in this paranoia about traitorous inner worlds and the tools that might make them visible. As Rebecca Lemov notes, in 1956, the sociologist Robert Merton had imagined a machine—the introspectometer—that would reveal and record everything people thought, felt, and saw: dreams, thoughts, and subjective experiences.[52] It was never built. But another fixture of Cold War fealty, the lie detector, provided a more modest (if unreliable) version of that dream. Deployed roughly fifty thousand times at Oak Ridge Tennessee between 1946 and 1953, it policed the nation's nuclear secrets and the political fidelity of the technicians who kept them.[53]

In their final fully developed version, the teaching machines of electronics maintenance became versions of the lie detector for Soucy (figure 5.6). Soucy claimed that maintenance staff often failed to report breakdowns as a way of covering up their own errors and those of their fellow crew members.[54] He was fond of citing the example of a missile technician who misconnected two adjacent (and identical) receptacles, producing a "scrubbed launch" and a damaged missile, but failed to report the mistake.[55] For Soucy, this general tendency of technicians to "scapegoat" electronic equipment, hiding untrustworthy humans behind fallible machines, meant that technicians were responsible for an even greater number of failures than anyone thought.[56] The teaching machines, Soucy hoped, would reveal those

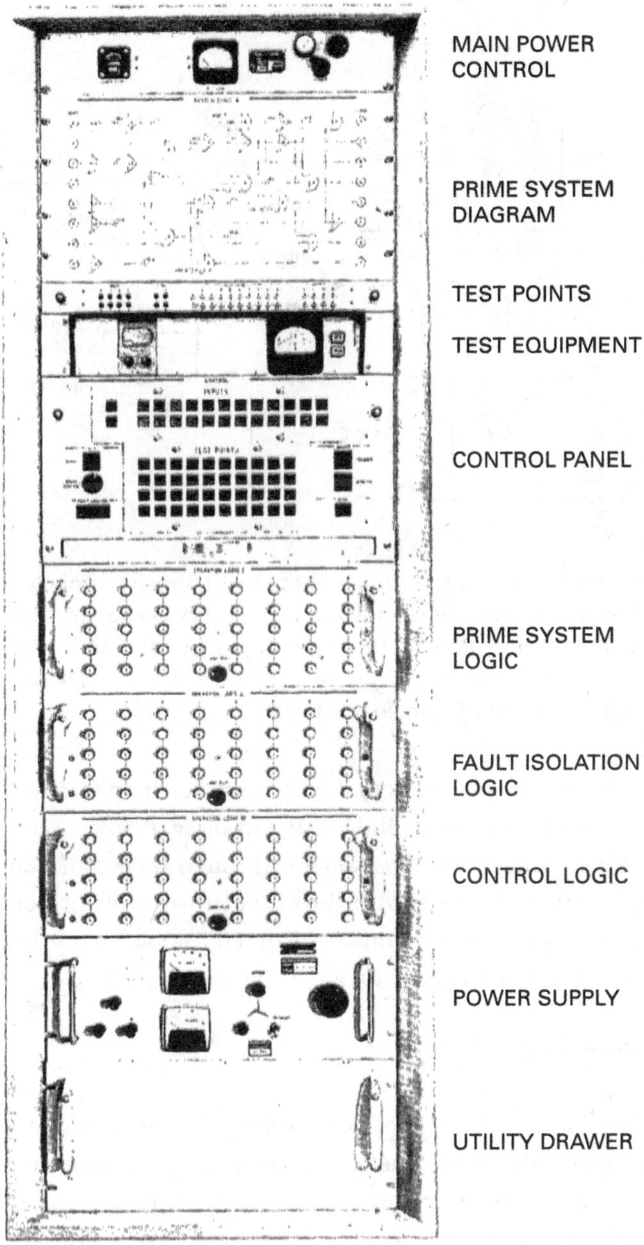

Figure 5.6

One of the more advanced teaching machines produced for military maintenance training: a six-foot relay rack whose reconfigurable circuitry could be used to "simulate" the data flow of a wide variety of systems. Researchers could "inject" faults by remotely interrupting signal flow. *Source:* Nicolai Buchaca and John Brinda Jr., "Maintainability Simulation and Demonstration Equipment Description and Application," *IEEE Transactions on Product Engineering and Production* 7 (1963), figure 2. Reprinted with permission from *IEEE Transactions on Product Engineering and Production 7* © 1963 IEEE.

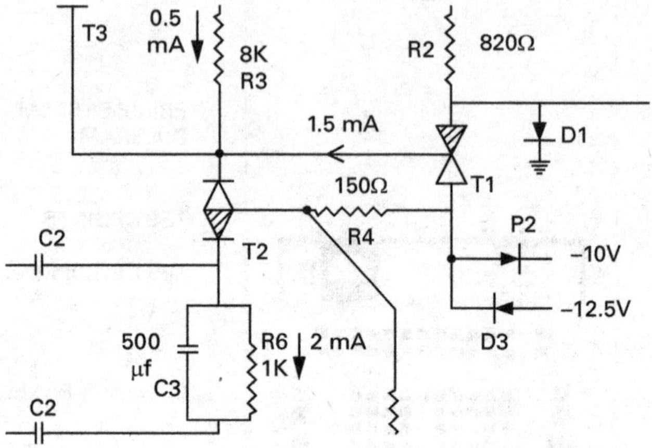

Figure 5.7
Detail from circuit diagram featuring nonstandard transistor symbols (T1 and T2) introduced at the Defence Research Board's electronics laboratory. *Source:* Norman Moody, circuit drawing, Alouette files, Canada Science and Technology Museum (CSTM) supplementary files that accompany the artifacts in the collection. Courtesy of the Canada Science and Technology Museum.

tendencies by measuring the actual propensity for human error. Although they made no claims to detect hidden agents or avert Soviet sabotage, their promise of discerning "real" breakdowns through the fog of human frailty and deception was of a piece with the other instruments of Cold War loyalty. And they would be accompanied by additional moral technologies targeting the other main group in Soucy's graph: designers.

Images of the Mind

When Soucy projected his graph in October 1958, it hinted at a world he could not freely talk about. The chasm at the graph's center represented a shadowy world of secret projects, undisclosed locations, secure facilities, and classified memos. In its invocation of "new world environments" and future technologies, though, the graph pointed toward the work of a single organization more than any other: the Canadian Defence Research Board (DRB). The DRB had been created in 1947 to coordinate Canadian military research in basic science and advanced technologies. It would soon elaborate some of the central doctrines of Cold War nuclear strategy, including the concepts of first-strike/second-strike planning, making it a center of

Cold War rationality alongside the RAND Corporation and other US think tanks of the period.[57] The DRB also led Canadian military research into the Cold War's "hostile environments." Its laboratories investigated equipment for arctic warfare, communications failures in high-northern latitudes, and electronics for supersonic flight and spaceborne operations. Those concerns wove the DRB's laboratories into signals intelligence projects for the National Security Agency (NSA) and advanced electronics projects at MIT's Lincoln Laboratory, placing it at the center of efforts to overcome the reliability crisis.[58]

In 1955, a curious symbol began appearing on circuit diagrams inside the main electronics research laboratories of the DRB. The symbol (see figure 5.7) was a schematic representation of the transistor, and it was the focus of a remarkable claim: that *drawing* transistors this way would help prevent breakdowns such as the ones fueling the reliability crisis.[59]

The claim seemed shocking on its face. How could a pencil-and-paper depiction of an electronic device ever influence the way it operated in the real-world applications of radar sets or guidance systems? And yet, the symbol was part of an explosion of competing schemes for transistors that made similar claims—at least fifteen public proposals and countless more private initiatives that swept across electronics journals, engineering schematics, sketch pads, and blackboards between 1947 and 1957.[60] To understand how Cold War technologists believed that drawings could avert aircraft malfunctions, radar breakdowns, and accidental nuclear launches, we have to make a brief detour through the history of electronic schematics and their relationship to inner worlds.

Diagrams are many things. Among their manifold functions and meanings, they are profoundly social texts. Part of that function stems from their role as proxies for bulky and unruly three-dimensional objects. By amassing, measuring, reproducing, and circulating diagrams, government agents, scientists, and social reformers (to name a few) exercise power over the material and social worlds these objects represent.[61] Partly because of this, diagrams have also historically defined both cultures of vision and communities of practice: from the technical drawings that bound eighteenth-century French military engineers in a common struggle against the "thick" objects of French artisans to the Feynman diagrams that fostered both a community of practitioners and a way of seeing within postwar theoretical physics.[62]

Early twentieth-century circuit diagrams represented particular ways of seeing, being, and belonging. They had originally developed from depictions of experimental electrical setups going back to the eighteenth century. But by the early twentieth century, they provided a shorthand for mathematical relationships that, through analogy, could apply to nonelectric phenomena such as sound.[63] Standing between thought and artifact, they increasingly acted as "paper tools"—sites where electronic devices were not only speculatively assembled, operated, taken apart, and reconfigured but also often first interrogated after breaking down.[64]

Under the massive expansion in electronics production in the 1950s, schematics increasingly coordinated a male-dominated world of design and manufacture, plagued by the growing possibilities of miscommunication and error. The problem stemmed from the confusion that postwar schematics produced in their audience. As the British physicist and radar researcher L. H. Bainbridge-Bell explained the process in 1950, in the halcyon days before the war, when the creation of electronic circuits had been the sole province of radio engineers, it had been difficult to draw confusing diagrams. The complex circuitry produced by wartime radar technologies, though, had made the production of schematics into a collaborative enterprise. Designers might initially produce a rough sketch of a circuit, with others adding to or altering the design as it developed, producing a confusing sketch that then moved to a draftsman for "cleaning up" and reproduction. These draftsmen "may have little knowledge of the functions of the circuit, or even of electricity."[65] Complicated circuits, combined with time pressures and lack of training for engineers, made confusing drawings a staple of early Cold War electronic design.

To make sense of these diagrams, Bainbridge-Bell explained, readers were often forced to draw their own "private" diagram. "Confusion has been known to arise when two people, each with his 'private' diagram, discuss a circuit over the telephone."[66] To help tame the potentially dangerous divergences and errors between these private mental images, Bainbridge-Bell suggested special conventions—"table manners," he called them—that expressed a growing conviction that the electronic arts were sustained and safeguarded by broader social, and explicitly moral, practices.[67] In the future debate over symbols, participants would repeatedly stress how electronic breakdowns emerged from violations of these wider duties of male-inflected courtesy, responsibility, and propriety.

The transistor represented a special case of the problem of "private" diagrams. Both the device and its symbol had been developed at Bell Laboratories in 1947 and envisioned as a replacement for vacuum tubes.[68] Very few of the circuit designers recruited in the thousands into Western militaries in the late 1940s and early 1950s had direct experience with transistors. In trying to understand what kinds of objects these new devices represented, engineers and circuit designers began developing their own symbols that they felt better embodied the nature and workings of the device.

The DRB symbol was part of this deluge of competing schemes. The electronics laboratory where the symbol emerged had reorganized itself to closely align with Soucy's view of breakdowns as a problem of environment, structure, and human error. Early in its history, the laboratory featured a Components Section that focused on developing robust individual electronic parts on the logic of Lusser's Law. But the growing tide of electronic failures, particularly in the new "revolutionary" transistors, inspired the laboratory to create a second section in 1954—Basic Circuits. The staff there devised transistor-specific design solutions that would both keep the devices within tight electrical parameters and make the circuits employing them relatively insensitive to variations in their performance—a perspective that also mirrored technical projects such as von Newmann's cellular automata or Paul Baran's work on packet switching, and social science projects such as Talcott Parsons's notion of subunits.[69] In this way, the social solutions that the Cold War "sciences of choice" proposed to tackle complexity mirrored the DRB's emerging ideas about the pathologies of machines, casting the problem of breakdown as a problem of organization, structure, system, and relation in the face of fallible constituent elements.[70]

Within that structure, the DRB symbol became a way of asserting a vision of electronic stability and social cooperation while preventing the design errors that generated electronic breakdowns. It had started as a "private symbol" developed by one of the laboratory's young recruits: Philip Thompson, a Cambridge physicist who had worked in telephony, electrical instruments for the RAF, and research tools for the Radio Section of the Cavendish Laboratory before joining DRB's Electronics Laboratory in 1954. It soon became a way of structuring the collaborative work of the lab, where a set of drawings might move from engineer to technician and back a dozen times, with individual circuits and elements contributed by as many people. Over the summer of 1957, Thompson's colleague, Jack Bateson, produced a memorandum that cast the symbol as part of a rationalized iconography—a "logical, self-consistent, graphical nomenclature"—for solid-state electronics, enrolling it in the creation of a community of interpreters and right thinkers designing the advanced electronics of the Cold War (figure 5.8).[71]

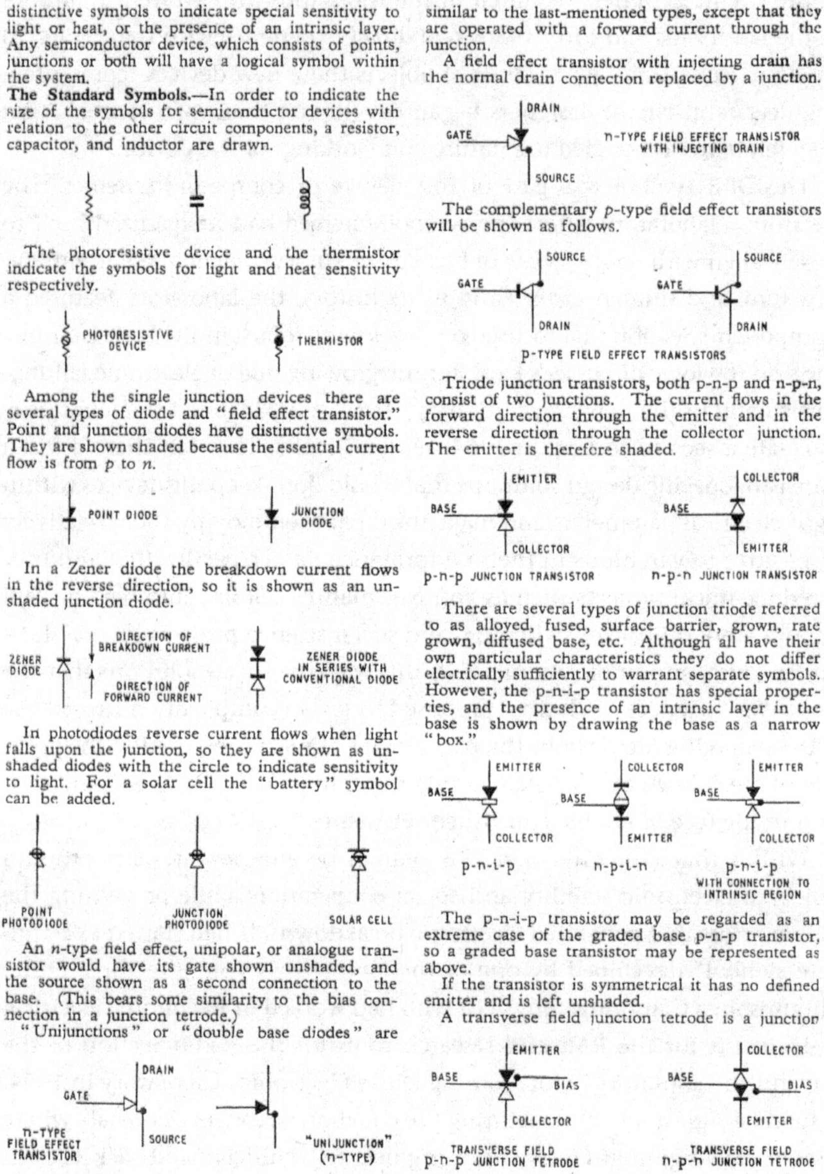

distinctive symbols to indicate special sensitivity to light or heat, or the presence of an intrinsic layer. Any semiconductor device, which consists of points, junctions or both will have a logical symbol within the system.

The Standard Symbols.—In order to indicate the size of the symbols for semiconductor devices with relation to the other circuit components, a resistor, capacitor, and inductor are drawn.

The photoresistive device and the thermistor indicate the symbols for light and heat sensitivity respectively.

PHOTORESISTIVE DEVICE
THERMISTOR

Among the single junction devices there are several types of diode and "field effect transistor." Point and junction diodes have distinctive symbols. They are shown shaded because the essential current flow is from *p* to *n*.

POINT DIODE
JUNCTION DIODE

In a Zener diode the breakdown current flows in the reverse direction, so it is shown as an unshaded junction diode.

ZENER DIODE
DIRECTION OF BREAKDOWN CURRENT
DIRECTION OF FORWARD CURRENT
ZENER DIODE IN SERIES WITH CONVENTIONAL DIODE

In photodiodes reverse current flows when light falls upon the junction, so they are shown as unshaded diodes with the circle to indicate sensitivity to light. For a solar cell the "battery" symbol can be added.

POINT PHOTODIODE
JUNCTION PHOTODIODE
SOLAR CELL

An *n*-type field effect, unipolar, or analogue transistor would have its gate shown unshaded, and the source shown as a second connection to the base. (This bears some similarity to the bias connection in a junction tetrode.)

"Unijunctions" or "double base diodes" are similar to the last-mentioned types, except that they are operated with a forward current through the junction.

DRAIN
GATE
SOURCE
n-TYPE FIELD EFFECT TRANSISTOR
"UNIJUNCTION" (n-TYPE)

A field effect transistor with injecting drain has the normal drain connection replaced by a junction.

DRAIN
GATE
SOURCE
n-TYPE FIELD EFFECT TRANSISTOR WITH INJECTING DRAIN

The complementary *p*-type field effect transistors will be shown as follows.

SOURCE
GATE
DRAIN
p-TYPE FIELD EFFECT TRANSISTORS

Triode junction transistors, both p-n-p and n-p-n, consist of two junctions. The current flows in the forward direction through the emitter and in the reverse direction through the collector junction. The emitter is therefore shaded.

EMITTER
BASE
COLLECTOR
p-n-p JUNCTION TRANSISTOR
n-p-n JUNCTION TRANSISTOR

There are several types of junction triode referred to as alloyed, fused, surface barrier, grown, rate grown, diffused base, etc. Although all have their own particular characteristics they do not differ electrically sufficiently to warrant separate symbols. However, the p-n-i-p transistor has special properties, and the presence of an intrinsic layer in the base is shown by drawing the base as a narrow "box."

EMITTER
BASE
COLLECTOR
p-n-i-p
n-p-i-n
p-n-i-p WITH CONNECTION TO INTRINSIC REGION

The p-n-i-p transistor may be regarded as an extreme case of the graded base p-n-p transistor, so a graded base transistor may be represented as above.

If the transistor is symmetrical it has no defined emitter and is left unshaded.

A transverse field junction tetrode is a junction

EMITTER
BASE
BIAS
COLLECTOR
TRANSVERSE FIELD p-n-p JUNCTION TETRODE
TRANSVERSE FIELD n-p-n JUNCTION TETRODE

Figure 5.8

The expanded diode-based system. Here, Thompson and Bateson elaborated both the system for constructing symbols and the rules for incorporating them into circuit diagrams. *Source:* P. M. Thompson and J. Bateson, "Semiconductor Symbols: Logical System for Diodes, Transistors and Other Junction Devices," *Wireless World* November 1957: 525–528. Reprinted by courtesy of *Electronic World*.

Under the growth of the reliability crisis, Thompson's supervisor, Norman Moody, turned this system into a way to discipline the mind of the circuit designer. As Soucy was formulating his ideas about the ultimate human causes of the reliability crisis, Moody used the podium at the 1957 Nuclear Instruments Conference at CERN—the high-energy physics laboratory straddling the French–Swiss border—to emphasize that Thompson's symbol was the key to avoiding precisely the kinds of design errors and moral lapses that populated Soucy's chart. Those errors most often stemmed, Moody believed, from a mistaken analogy between vacuum tubes and transistors, reinforced by the original Bell symbol for the transistor. "To designers who are familiar with vacuum tubes," which was most anyone working in military electronics, Moody explained, "this trivial resemblance [of power amplification] . . . has been a trap. Many of their failures are strewn along the path of progress."[72] Designers had to approach transistors fresh, learning entirely new circuit techniques, avoiding tube-by-tube replacement by transistors, redesigning existing circuits on a functional basis, and ultimately spending at least a year laboring with the new devices in order to command their techniques.[73] Avoiding electronic failures was therefore a problem of making engineers and circuit designers trustworthy by making them *think* about the transistor in more reliable ways, avoiding the bypaths leading away from "the haven of safe, and reliable design."[74]

For Moody's group, these new symbols ultimately proved how material complexity, individual weakness, and overall stability might coexist, despite the tirades from Redstone Arsenal. In designing a digital computer for use in the DRTE, for example, Moody effused over a "family of circuits that thrives on heavy load currents, often to the limit set by transistor destruction." Singling out one circuit from their family of configurations, Moody explained that "though relatively complex in itself, [it] yields an excellent performance in the face of wide transistor and component variations."[75] The DRB designs would soon enter the "new world environments" Soucy envisioned through radar systems for Canadian supersonic aircraft and electronic pulse circuits for an ionospheric satellite. Despite those developments, though, by the early 1960s, the competing transistor icons, like teaching machines, had been largely abandoned. For the brief period when they flourished, however, these drawings interwove the inner worlds of Cold War circuit designers, the social worlds they occupied, and the material worlds they were responsible for creating. They structured both

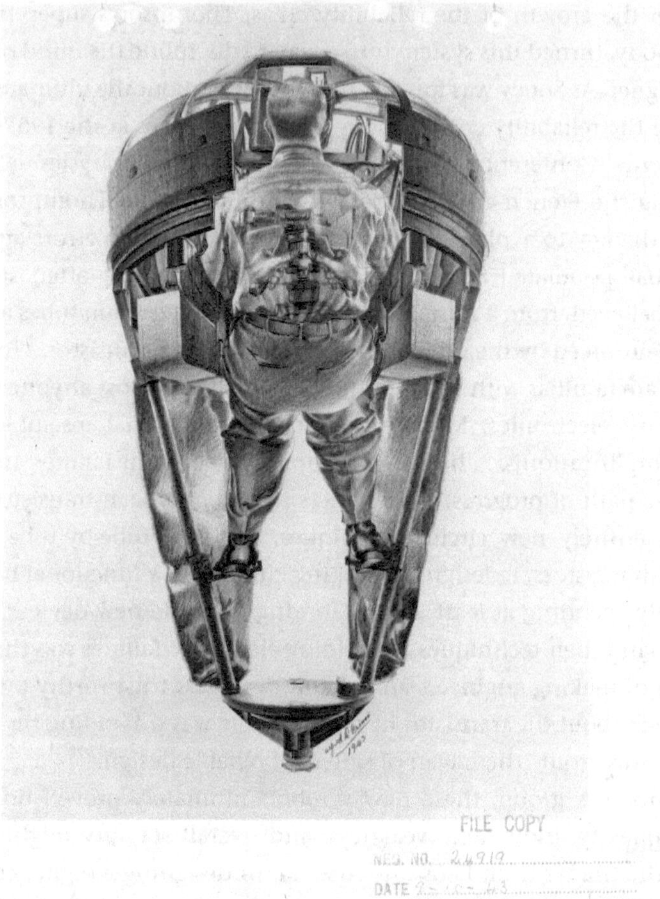

Figure 5.9
Alfred Crimi, drawing of operator inside Sperry gun turret. *Credit:* A. D. Crimi sketches, 1943. Box 149, folder 3, AVD_1986273_149_03_24919, Sperry Gyroscope Company Division photographs and films, accession 1986-273-V, Hagley Museum & Library, Wilmington, DE, 19807. Courtesy of Hagley Museum and Library.

the electrical environments that would keep electronic parts safe and the conceptual and social worlds that Soucy and others hoped would guide the thoughts of the West's technical staff and avert the breakdowns that had made them necessary in the first place.

Conclusion

Almost two decades before Soucy projected his chart of the reliability crisis, another image had illuminated the concerns around "universal" humans, advanced weapons, and mid-century visions of global war. Sometime in 1941, as World War II entered its most desperate phase for the Allies, an émigré Italian fresco and mural painter working for Sperry Gyroscope Company sketched a vision of a gun turret for a US B-Series bomber (figure 5.9). Alfred Crimi's illustration was most striking for the way it made the gunner's body selectively transparent, so that mechanical controls seemed to take the place of vital organs.[76] Even as the image signaled the fading supremacy of the mechanical, its integration of the human and the machinic indicated a central concern of the reliability crisis to come: how to disentangle the failings of humans from faults of machines when things broke down.

This image of the cyborg, the cybernetic organism that melded human and machine, has dominated histories of Cold War technology alongside accounts of the silicon revolution in electronics. Rooted in the person of the operator and the faith that automated machines could overcome human limitations, cybernetics tackled the delicate balancing act of dividing labor and responsibility between people and technologies in ways that maximized the strengths of each. But Crimi's image was also a metaphor for everything that the dreams of automation hid. As the reliability crisis made clear, automation *generated* labor. Endless psychological studies, strategic assessments, personnel evaluations, teaching machines, training regimes, preparedness analyses, reformed drawing practices, and restructured laboratories testified to the labor it produced. This paradox of Cold War automation—imagined as labor-saving yet generating new, widespread, intensive, and hidden forms of labor—lay at the heart of the reliability crisis and still haunts the technological fantasies of Silicon Valley today.[77]

Here and throughout the episodes covered in this book, solutions to the problems of technology have been solutions to the problem of social order.[78] By giving breakdowns the form and language of crisis, Soucy and his contemporaries highlighted how these events disrupted Western strategies of global dominance and threatened the survival of Western personnel and machines and the political and ideological systems they supported. Gathering those events in graphs, memos, and reports, artificially generating breakdowns and placing them at the core of psychological studies and new teaching technologies, they transformed breakdowns into productive disturbances: events designed to immediately elicit a protocoled response that stabilized ideas about the Cold War social order taking shape around them. Confronting problems of complexity that resonated with the challenges of liberal democracies and the trials of modern society, their language and concerns both drew from and authorized the work of social scientists, policy experts, and Cold War intellectuals around a common solution: a moral and disciplined technical self, supported by systems for right thinking, and envisioned as essential to both the functioning of complex technologies and the stability of Western democracies.

The reliability crisis, though, teaches us that indulging the technological fantasy of automation was possible only as long as one shared the tight framing of Crimi's image. Outside it—in the world of Soucy's graph, Thompson's transistor symbols, and the military's teaching machines—other kinds of people and experiences of technology mattered. Crimi's tranquil image pushed those other selves to the very edges of its vision, along with the supreme violence that sustained it. But the tranquility of the image also pointed to something less obvious, something that had haunted Soucy's graph and the histories of cybernetics as well: the nearly punctiform instant always just before—just before the crumpling of metal, the jamming of gears, the panic of the operator, the intervention of the technician, and the blame of the designer. It was the moment just before the only moment Cold War officials knew would arrive for certain: the moment when things began to fall apart.

Epilogue: Beyond Heidegger's Hut

Sometime in the mid-1920s, in a hut on the edge of the Black Forest—so the story goes—breakdowns became portals into worlds we thought we knew. Heidegger had been coming to this small three-room cabin, his "work-world," in the hills above Todtnauberg since 1922, composing lectures and working out ideas that would shape his famously impenetrable treatise, *Being and Time*. Breakdowns filled the cities and spreading industrial landscapes of the fledgling Weimar Republic, the liberal democratic regime that emerged from Germany's humiliating defeat in World War I. Short circuits, power failures, breakages, and malfunctions ripped through the modern machinery of Berlin, Hamburg, Munich, and Frankfurt. But these were *not* the events that interested Heidegger.

After all, Heidegger's hut was a mountain refuge—an escape from the modern industrial technologies of his age, the ones Heidegger would increasingly deride after World War II, and from the liberal politics of Weimar, which would cease to exist when the Nazi's seized power in 1933. Instead, the breakdowns that darkened the doorway of his cabin were the imagined breakages of *premodern things*: the timeless tools of the artisan and the peasant. In these thick, resonant objects, breakage was a moment of revelation, a portal into another dimension of experience that showed their otherwise "transparent" backgrounds of use, practices, actions, and relations.[1]

For all its talk of revelation, though, Heidegger's vision hid at least as much as it revealed. First, it obscured its own politics. The archaic technologies he pressed into the service of revelation were no innocent philosophical props. Heidegger knew how they hailed a mythical German past rooted in romantic nationalism and its racist conceptions of *volk* and social order.[2] Second, his philosophical vision hid a historical alternative that had long been circulating beyond his mountain hut. For almost two centuries, a related politics of breakdowns had been building in the nations that had recently waged war against Germany and that would do so again. Where Heidegger's vision was anti-modern, this other politics had been proudly of the moment. Where Heidegger focused on the premodern tools of the peasant and the artisan, this politics focused on the latest innovations in state violence, mass transportation, industrial-scale agriculture, mass manufacturing, and modern warfare. Where Heidegger looked to a conservative, mythic, and idealized past, this politics proffered visions of liberal society, technological progress, and ideal social futures. And where Heidegger saw breakage as a portal into the timeless nature of being, this alternative saw in breakdowns both a threat and a solution to contemporary problems of social order, technology, and the modern soul.

This book has drawn its stories from that other history. It has explored how medical officials, engineers, politicians, consultants, and social reformers turned indifference to breakdowns into a problem of the self—a problem of how to make broken machines into disturbances for the people who experienced them. Those figures premised their projects on a belief that the kinds of people who mattered to the social order were also the kinds of people who would be troubled by the broken machine. Across different places and times, they worked to shape versions of an "us" that breakdowns helped distinguish: a sentimental self for the French Revolution, a responsible self for Victorian Britain, an efficient self for Machine Age America, or a rational and reasoning self for the North American Cold War. Where they saw that quality lacking, they developed techniques to cultivate, nurture, and enforce it in some and to cast it as an inherent deficiency in others. In doing so, they turned breakdowns into moments that measured artisans, railway workers, enslaved people, factory hands, circuit designers, and technicians against cultural competencies, cultivated habits, and psychological dispositions that defined who belonged to the social orders of their times and who stood outside them—who counted as "us."[3] The fact that others felt differently—ignoring breakdowns or celebrating them, seeking rest rather than revelation in the pauses they provided—was precisely what gave those moments their diagnostic value.[4] These moments showed how broken machines might extend the social order and its distinctions rather than suspend them. Through technologies of punishment, transportation, production, and war, these efforts gave shape to cosmologies of modern technologies and constellations of the self that persist into the twenty-first century.

These histories are not gone. Some of our most influential theories of technology—the broken world and the working world—are part of their legacy. Any broken tool is "disturbing to us," Heidegger claimed. If anything made that assertion both possible and plausible in the 1920s, it was an unacknowledged cultural history rather than a transhistorical truth. Breakage was always much more than a moment of philosophical clarity. In the 1920s, in the hands of one of the preeminent philosophers of the time, it was a moment to hail an imagined community—an "us" that was knowingly and devastatingly divisive—with all the social and political power that entailed. An ocean away in the United States, working in that same turbulent decade, Lewis Mumford imagined a different community of breakage and the self. Working out many of the ideas that would inform his own monumental work, *Technics and Civilization*, Mumford surveyed the middle decades of nineteenth-century America and identified a tragedy. The Civil War—a war fought over the destruction of slavery—represented for Mumford the dark dividing line between a Golden Day of American individualism and the tragic triumph of the machine.[5] The new "machine-trained populations" that resulted were precisely the ones, he would lament, who were indifferent to breaking technologies.[6] Lamenting the rise of the machine, the death of American individualism, and the insensitivity to breakdown was of a piece for Mumford. Heidegger and the broken world; Mumford and the working world. Both lamented the consequences of industrial modernity; both believed disturbance was the true and authentic response to broken machines. Their perspectives were aligned with many of the figures in this book—people working, across centuries, against the tendencies Mumford lamented and in service of the disturbances Heidegger hailed.

But there were always other ways of living among broken machines. In the dancers of the breakdown, in the steam men of the Mississippi, and in the workers of the gins of South Carolina and the Georgia Sea Islands, other cosmologies arose. And they were not alone. The geographies of belonging that this book has traced, and that continue to be drawn around breakdowns, have always been filled with other visions. They push us to recognize that breakdowns *are* portals into worlds we thought we knew—worlds about which we imagined not much more could be said. But, against both Heidegger and Mumford, they are portals into the historical and contemporary politics of the broken machine and the related histories of the self. Heidegger, seized by the power of breakage and the prospects of his insights, hid this history. Mumford, steeped in his view of American decline, could not see that he was a part of it. Those moments of breakage reveal not only how people have historically built certain selves into the workings of machines but also how they have built selves around their malfunctions and silences. Every indication is that we, their inheritors, will continue to do so for the machines to come.

Why have these stories been largely absent from the history of technology? It's tempting to call this a modern tendency—part of a history of forgetting. Like disasters, breakdowns are put to specific uses and then cast aside. But forgetting is too easy and comforting an explanation. It ignores how forgetting is also an act. Histories of breakage are, not coincidentally, histories of the things we throw away. Forgetting them means engaging in the practices that Trouillot believed created the silences of our past: the historical acts that valorize and diminish, the archival practices that preserve and discard, the historical methods that acknowledge and ignore.[7] One of the arguments of this book has been that histories of breakdown are not histories of the useless. The material and social worlds they described deserve attempts to remember and preserve. Making them strange again unsettles the condescending relations they have helped construct. And it points, as Steven Jackson, suggests, to the possibilities of building "new and different forms of solidarity with our objects," not as a return to some lost age of communion with machines that Heidegger or even Mumford lamented but rather as a way of understanding our collective present.[8]

As Arjun Appadurai and Neta Alexander have observed for our own moment, "It is surprisingly common to trivialize, ignore, or simply forget the numerous ways in which technology breaks, malfunctions, or frustrates its users. Of course this 'forgetfulness' . . . attests not to the failure of memory, but to the success of cultural fantasies and collective imaginaries of technological utopias."[9] They note the contemporary practices that aid this forgetting: diverting software glitches into bug reports, rebooting stalled Wi-Fi routers, or accepting the planned obsolescence of our technologies. Less obviously, contemporary attitudes merge, as they did for Heidegger and Mumford, with scholarly practices: We ignore or trivialize or forget breakdowns, channeling them into something less threatening, moving quickly past them.[10] Along the way, we make our institutions more durable, and we close down questions about our full relationships to machines before they become too uncomfortable.

"When and to what end do we tell ourselves stories of failure?" In asking that question, Appadurai and Alexander implore us to remember the failures of the recent past as a way of recapturing what is most urgent about our present.[11] This book has argued that stand-alone stories and the recent past are not enough. Fully understanding what failures do in our present means coming to terms with the longer histories of our most cherished assumptions—what Stuart Hall calls "the horizon of the taken-for-granted" and Lorraine Daston calls "histories of the self-evident."[12] Like the natural disasters that also disappear too quickly from modern memory, the disappearance of breakdowns says something about the intimate relationships crafted around them and with them. And it raises a question for another time: What does it mean to forget when we are surrounded by reminders?

Notes

Introduction

1. Bruno Latour, "Technology Is Society Made Durable," *Sociology Review* 38, no. 1 (1990): 103–131.

2. For the histories of the guillotine, see, e.g., Daniel Arasse, *The Guillotine and the Terror* (New York: Penguin, 1989); Carol, *Physiologie de la veuve* (Seyssel, France: Champ Vallon, 2012); Alister Kershaw, *A History of the Guillotine* (London: John Calder, 1958); Dorinda Outram, *The Body and the French Revolution: Sex, Class and Political Culture* (New Haven, CT: Yale University Press, 1989), 106-123; Nina Athanassoglou-Kallmyer, "Gericault's Severed Heads and Limbs: The Politics and Aesthetics of the Scaffold," *The Art Bulletin* 74, no. 4 (1992): 599–618; Georges Canguilhem, "Le Cerveau et la pensée," in *Georges Canguilhem: Philosophe, historien des sciences* (Paris: Albin Michel, 1993), 11–33; Michel Foucault, *Discipline and Punish: The Birth of the Prison*, trans. Alan Sheridan (New York: Vintage, 1995); Regina Janes, "Beheadings," *Representations* 35, no. 1 (1991): 21–51; Ludmilla Jordanova, "Medical Mediations: Mind, Body and the Guillotine," *History Workshop* 28, no. 1 (1989): 39–52; and Jürgen Martschukat, "Ein Schneller Schnitt, Ein Sanfter Tod: Die Guillotine Als Symbol Der Aufklärung," in *Das Volk Im Visier Der Aufklärung: Studien Zur Popularisierung Der Aufklärung Im Späten 18. Jahrhundert*, ed. Anne Conrad, Arno Herzig, and Franklin Koptisch (Hamburg: Lit, 1998), 121–142. For the general emphasis on working in history of technology and technology studies, see footnote 13.

3. On the dual instability of social and material forms, see, e.g., John Law, "Technology and Heterogeneous Engineering: The Case of Portuguese Expansion," in *The Social Construction of Technological Systems: New Directions in the Sociology and History of Technology*, ed. Wiebe E. Bijker, Thomas P. Hughes, and Trevor Pinch (Cambridge, MA: MIT Press, 2012), 105–127.

4. See, e.g., Simon Schaffer, "Easily Cracked: Scientific Instruments in States of Disrepair," *Isis* 102, no. 4 (2011): 706–717; Bruno Latour, *Aramis, or, The Love of Technology* (Cambridge, Mass: Harvard University Press, 1996) and *Reassembling the Social:*

An Introduction to Actor-Network-Theory (New York: Oxford University Press, 2005), 81; Susan Leigh Star, "The Ethnography of Infrastructure," *American Behavioral Scientist* 43, no. 3 (1999): 377–391 and "Power, Technology and the Phenomenology of Conventions: On Being Allergic to Onions," in *A Sociology of Monsters*, ed. John Law (London: Routledge, 1991); and David Edgerton, *The Shock of the Old: Technology and Global History Since 1900* (Oxford: Oxford University Press, 2011), chap. 4.

5. Schaffer, "Easily Cracked."; Edgerton, *The Shock of the Old*, chap. 4; Steven J. Jackson, "Rethinking Repair," in *Media Technologies: Essays on Communication, Materiality, and Society*, ed. Tarleton Gillespie, Pablo J. Boczkowski, and Kirsten A. Foot (Cambridge, MA: MIT Press, 2014); Stephen Graham and Nigel Thrift, "Out of Order: Understanding Repair and Maintenance," *Theory, Culture and Society* 24, no. 3 (2007): 1–25; and Andrew Russell and Lee Vinsel, "After Innovation, Turn to Maintenance," *Technology and Culture* 59, no. 1 (2018): 1–25. For a review of the literature on maintenance, and its neglect, see David J. Denis, "Why Do Maintenance and Repair Matter," in *The Routledge Companion to Actor-Network Theory*, eds. Anders Blok, Ignacio Farias, and Celia Roberts, (New York: Routledge, 2019), 283–93. See also Julian E. Orr, *Talking About Machines: An Ethnography of a Modern Job* (Ithaca, NY: Cornell University Press, 1996), 79; Frank Nutch, "Gadgets, Gizmos, and Instruments: Science for the Tinkering," *Science, Technology, and Human Values* 21 (1996): 214–228, esp. 215; and Christopher Henke, "The Mechanics of Workplace Order: Toward a Sociology of Repair," *Berkeley Journal of Sociology* 44 (2000): 55–81, esp. 61.

6. On invisible technicians, see Steven Shapin, "The Invisible Technician," *American Scientist* 77 (1989): 554–563. For a survey of the role of technicians in histories of science and technology, see Iwan Rhys Morus, "Invisible Technicians, Instrument-Makers and Artisans," in *A Companion to the History of Science*, ed. Bernard Lightman (London: John Wiley, 2016), 97–110.

7. This view of breakdowns as revelation draws heavily on Heidegger's famous discussion of breakage in *Being and Time*. See Martin Heidegger, *Sein Und Seit* (Tübingen, Germany: Max Niemeyer Verlag, 1967), 74. For works following this tradition, see, e.g., Peter Galison, "On the Building, Crashing, and Thinking of Technologies and Selfhood," in *Art in the Anthropocene: Encounters Among Aesthetics, Politics, Environments, and Epistemologies*, ed. Etienne Turpin and Heather Davies (London: Open Humanities Press, 2015), 181–190; Peter Galison, "An Accident of History," in *Atmospheric Flight in the Twentieth Century*, ed. Peter Galison and Alex Roland (Dordrecht, Netherlands: Springer, 2000); Leigh Star, "The Ethnography of Infrastructure."; Latour, *Reassembling the Social*; Peter-Paul Verbeek, *What Things Do: Philosophical Reflections on Technology, Agency, and Design* (University Park: Pennsylvania State University Press, 2005), 79; Graham and Thrift, "Out of Order."; Jackson, "Rethinking Repair."; and Neta Alexander, "Rage Against the Machine: Buffering, Noise, and Perpetual Anxiety in the Age of Connected Viewing," *Cinema Journal* 56, no. 2 (2017): 14. For a brilliant critique of this view from the perspective of African studies, see Jenna

Burrell, *Invisible Users: Youth in the Internet Cafés of Urban Ghana* (Cambridge, MA: MIT Press, 2012), 14. As Burrell notes, the view of breakdown as unusual is critical for the analytical force that breakdowns hold for mainstream science and technology studies, which naturalizes stable functioning and portrays breakdowns as "out of the ordinary, . . . sudden and transitory." For an example, see Latour, *Reassembling the Social*, 81.

8. Jackson, "Rethinking Repair," 221.

9. Steven Shapin and Simon Schaffer, *Leviathan and the Air-Pump: Hobbes, Boyle, and the Experimental Life* (Princeton, NJ: Princeton University Press, 1985), 332.

10. Nella Larsen, *Quicksand* (New York: Alfred A. Knopf, 1928), 18. Saidiya Hartman builds on Larsen's "disturbing factor" in *Wayward Lives, Beautiful Experiments: Intimate Histories of Riotous Black Girls, Troublesome Women, and Queer Radicals* (New York: Norton & Company, 2019).

11. Naomi Klein, *The Shock Doctrine: The Rise of Disaster Capitalism* (Toronto: Knopf Canada, 2007); and Daniel Immerwahr, "All That is Solid Bursts into Flame: Capitalism and Fire in the Nineteenth-Century United States," *Past and Present* 265, no. 1 (2024): 97–138. On technological disasters, see also Scott Gabriel Knowles, *The Disaster Experts: Mastering Risk in Modern America* (Philadelphia: University of Pennsylvania Press, 2013); Sara B. Pritchard, "An Envirotechnical Disaster: Nature, Technology, and Politics at Fukushima," *Environmental History* 17, no. 2 (2012): 219–243; Adriana Petryna, *Life Exposed: Biological Citizens after Chernobyl* (Princeton, NJ: Princeton University Press, 2013); Peter Soppelsa, "Paris's 1900 Universal Exposition and the Politics of Urban Disaster," *French Historical Studies* 36, no. 2 (2013): 271–298; and Kim Fortun, *Advocacy after Bhopal: Environmentalism, Disaster, New Global Orders* (Chicago: University of Chicago Press, 2001).

12. Schaffer, "Easily Cracked," 709.

13. For the emphasis on working, see, e.g., Lewis Mumford, *Technics and Civilization* (New York: Harcourt Brace Jovanovich, 1934); Sigfried Giedion, *Mechanization Takes Command: A Contribution to Anonymous History* (New York: Oxford University Press, 1948); Thomas P. Hughes, *Networks of Power: Electrification in Western Society, 1880–1930* (Baltimore, MD: Johns Hopkins University Press, 1983); Wiebe E. Bijker and John Law, eds., *Shaping Technology / Building Society: Studies in Sociotechnical Change* (Cambridge, MA: MIT Press, 1992); Donald A. MacKenzie and Judy Wajcman, *The Social Shaping of Technology* (Cambridge, MA: MIT Press, 1999); Lucy A. Suchman, *Human–Machine Reconfigurations: Plans and Situated Actions* (New York: Cambridge University Press, 2007); H. M. Collins, "The Tea Set: Tacit Knowledge and Scientific Networks," *Science Studies* 4, no. 2 (1974): 165–185; Steven Shapin and Simon Schaffer, *Leviathan and the Air-Pump: Hobbes, Boyle, and the Experimental Life* (Princeton, NJ: Princeton University Press, 1985); Bruno Latour, "Give Me a Laboratory and I Will Raise the World," in *Science Observed: Perspectives on the Social Study of Science*,

ed. Karin Knorr-Cetina and Michael Mulkay (London: Sage, 1983), 141–170; and Wiebe E. Bijker et al., eds., *The Social Construction of Technological Systems: New Directions in the Sociology and History of Technology* (Cambridge, MA: MIT Press, 2012).

14. Notable exceptions to the neglect of breakdowns as historical objects are Schaffer, "Easily Cracked."; John G. Burke, "Bursting Boilers and the Federal Power," *Technology and Culture* 7, no. 1 (1966): 1–23; and Yuan Yi, "Malfunctioning Machinery: The Global Making of Chinese Cotton Mills, 1877-1937," (PhD diss., Columbia University, 2020).

15. Michel-Rolph Trouillot, *Silencing the Past: Power and the Production of History* (Boston, MA: Beacon Press, 2015), 96 and chap. 3.

16. On the neglect of accidents in history, see Roger Cooter and Bill Luckin, eds., "Accidents in History: An Introduction," in *Accidents in History: Injuries, Fatalities, and Social Relations* (Amsterdam: Rodopi, 1997), 5. On histories of accidents, see K. Figlio, "What Is an Accident?" in *The Social History of Occupational Health*, ed. Paul Weindling (London: Croom Helm, 1985), 180–206; Paul Fyfe, *By Accident or Design: Writing the Victorian Metropolis* (New York: Oxford University Press, 2015), 256; Judith Green, *Risk and Misfortune: A Social Construction of Accidents* (London: UCL Press, 1997); and Ryan Vieira, "The Epistemology and Politics of the Accidental: Connecting the Accident's Intellectual and Cultural Historiography," *History Compass* 11 (2013): 227–234. On technological accidents, see Wolfgang Schivelbusch, *The Railway Journey: The Industrialization of Time and Space in the Nineteenth Century* (Berkeley: University of California Press, 2014); Charles Perrow, *Normal Accidents: Living with High-Risk Technologies* (Princeton, NJ: Princeton University Press, 1999); Jamie L. Bronstein, *Caught in the Machinery: Workplace Accidents and Injured Workers in Nineteenth-Century Britain* (Stanford, CA: Stanford University Press, 2008); Gabrielle Hecht, "Enacting Cultural Identity: Risk and Ritual in the French Nuclear Workplace," *Journal of Contemporary History* 32, no. 4 (1997): 483–507; John Burnham, *Accident Prone: A History of Technology, Psychology and Misfits of the Machine Age* (Chicago: University of Chicago Press, 2009); and M. Esbester and J. Wetmore, eds., "(Auto) Mobility, Accidents, and Danger," special issue of *Technology and Culture* 52 (2015).

17. Mumford, *Technics and Civilization*, 13–14.

18. Mumford, *Technics and Civilization*, 361. Mumford linked this separation to the machine-fueled rise of objectivity in the nineteenth century. Mumford singled out "rural populations" as exceptions to this indifference.

19. Cruikshank argues that these technologies are endemic in liberal democracies because they act at the limit of individual freedoms, encouraging behaviors that are otherwise impossible to enforce. Barbara Cruikshank, *The Will to Empower* (Ithaca: Cornell University Press, 1999), 1-2.

20. Schaffer, "Easily Cracked," 707.

21. On Barak, *On Time: Technology and Temporality in Modern Egypt* (Berkeley: University of California Press, 2013), 63. See also Simon Schaffer, "Late Victorian Metrology and Its Instrumentation: A Manufactory of Ohms," in *The Science Studies Reader*, ed. Mario Biagioli (New York: Routledge, 1999); and Janet Abbate, *Inventing the Internet* (Cambridge, MA: MIT Press, 2000).

22. Edward Jones-Imhotep, "Introduction," in *The Unreliable Nation: Hostile Nature and Technological Failure in the Cold War* (Cambridge, MA: MIT Press, 2017); Gregg Mitman, "Where Ecology, Nature, and Politics Meet: Reclaiming the Death of Nature," *Isis* 97, no. 3 (2006): 496–504; Keith Wailoo et al., eds., *Katrina's Imprint: Race and Vulnerability in America* (New Brunswick, NJ: Rutgers University Press, 2010); Brian Larkin, *Signal and Noise: Media, Infrastructure, and Urban Culture in Nigeria* (Durham, NC: Duke University Press, 2008); and Pritchard, "An Envirotechnical Disaster."

23. Jackson, "Rethinking Repair."

24. Burke, "Bursting Boilers and the Federal Power"; Perrow, *Normal Accidents*; Ulrich Beck, *Risk Society: Towards a New Modernity*, trans. Mark Ritter (Thousand Oaks, CA: Sage, 1992); Diane Vaughan, *The Challenger Launch Decision: Risky Technology, Culture, and Deviance at NASA* (Chicago: University of Chicago Press, 1996); Tom Crook and Mike Esbester, eds. *Governing Risks in Modern Britain: Danger, Safety and Accidents, c. 1800–2000* (London: Palgrave Macmillan, 2016); Green, *Risk and Misfortune*; Scott Knowles, *The Disaster Experts*; Arwen Mohun, *Risk: Negotiating Safety in American Society* (Baltimore: Johns Hopkins University Press, 2012); and Rebecca Slayton, *Arguments That Count: Physics, Computing, and Missile Defense, 1949–2012* (Cambridge, MA: MIT Press, 2013).

25. See, e.g., Jane Bennett, *Vibrant Matter* (Durham, NC: Duke University Press, 2009); Bruno Latour, "Where Are the Missing Masses? The Sociology of Mundane Artifacts," in *Shaping Technology / Building Society: Studies in Sociotechnical Change*, ed. Wiebe E. Bijker and John Law (Cambridge, MA: MIT Press, 1992), 151; and Bruno Latour, "When Things Strike Back: A Possible Contribution of 'Science Studies' to the Social Sciences," *British Journal of Sociology* 51 (2020): 107–123.

26. Several scholars, including Moore, have questioned the newness of new materialism. See, e.g., Sara Ahmed, *Queer Phenomenology: Orientations, Objects, Others* (Durham, NC: Duke University Press, 2006); and Mukharji, "Occulted Materialities." On the limits of new materialism see Kyla Wazana Tompkins, "On the Limits and Promise of New Materialist Philosophy," *Emergent Critical Analytics for Alternative Humanities* 5.1 (Spring 2016). See also Angela Willey, "A World of Materialisms: Postcolonial Feminist Science Studies and the New Natural," *Science, Technology, & Human Values* 41, no. 6 (2016): 991–1014.

27. For the case of "danno" and "ruina" in early modern Italy, for example, see Francis C. Moon, *The Machines of Leonardo Da Vinci and Franz Reuleaux: Kinematics of Machines From the Renaissance to the 20th Century* (Dordrecht, Netherlands: Springer,

2007), 38; and Pamela O. Long, *Engineering the Eternal City: Infrastructure, Topography, and the Culture of Knowledge in Late Sixteenth-Century Rome* (Chicago: University of Chicago Press, 2018), 93.

28. Despite being widespread, machine failures only began to be discussed widely in printed literature in the mid-nineteenth century. See Moon, *The Machines of Leonardo Da Vinci and Franz Reuleaux*.

29. Cooter and Luckin, *Accidents in History*, 5.

30. Galison, "An Accident of History."

31. Stephen Greenblatt, *Renaissance Self-Fashioning: From More to Shakespeare* (Chicago: University of Chicago Press, 2005), 1-3. For a recent overview of the history of the self, see Jerrold Seigel, *The Idea of the Self: Thought and Experience in Western Europe Since the Seventeenth Century* (New York: Cambridge University Press, 2005).

32. Eric Schatzberg's recent examination, for example, excludes the self from the definition of technology. See Eric Schatzberg, *Technology: Critical History of a Concept* (Chicago: University of Chicago Press, 2018), 2. For histories of technology that take up the question of the self, see, e.g., Mumford, *Technics and Civilization*; Schivelbusch, *Railway Journey*; Bray, *Technology, Gender and History in Imperial China*; and David E. Nye, *Technology Matters: Questions to Live With* (Cambridge, MA: MIT Press, 2006). On historical and philosophical investigations that engage the self, see, e.g., Foucault, *Discipline and Punish*; Ian Hacking, *Rewriting the Soul: Multiple Personality and the Sciences of Memory* (Princeton, NJ: Princeton University Press, 2003); Lorraine Daston and Peter Galison, *Objectivity* (New York: Zone Books, 2007). For a detailed discussion of Foucault and technology, see Michael C. Behrent, "Foucault and Technology," *History and Technology* 29, no. 1 (2013): 54–104. On normative accounts of selfhood, see Arnold Davidson, "Spiritual Exercises and Ancient Philosophy: An Introduction to Pierre Hadot," *Critical Inquiry* 16, no. 3 (1990): 475–482; Pierre Hadot, *Philosophy as a Way of Life: Spiritual Exercises from Socrates to Foucault*, ed. Arnold I. Davidson (Malden, MA: Blackwell, 1995). On the self as an object of governance, see, e.g., Patrick Joyce, *The Rule of Freedom: Liberalism and the Modern City* (New York: Verso, 2003); and Nikolas Rose, *Governing the Soul: The Shaping of the Private Self* (Stanmore, UK: Free Publishing, 1999).

33. Daston and Galison, *Objectivity*.

34. Latour, "Where Are the Missing Masses?"

35. See Judy Wajcman, *Technofeminism* (New York: Polity, 2004); Francesca Bray, *Technology, Gender and History in Imperial China* (New York: Routledge, 2013), "Gender and Technology," *Annual Review of Anthropology* 36, no. 1 (2007): 37–53; and *Technology and Gender: Fabrics of Power in Late Imperial China* (Berkeley: University of California Press, 1997); Ruth Schwartz Cowan, *More Work for Mother: The Ironies of Household*

Technology from the Open Hearth to the Microwave (New York: Basic Books, 1983); Nina E. Lerman, Arwen P. Mohun, and Ruth Oldenziel, "Versatile Tools: Gender Analysis and the History of Technology," *Technology and Culture* 38, no. 1 (1997): 1–8; Maria Lohan and Wendy Faulkner, "Masculinities and Technologies: Some Introductory Remarks," *Men and Masculinities* 6 (2004): 319–329; Nina E. Lerman, "Categories of Difference, Categories of Power: Bringing Gender and Race to the History of Technology," *Technology and Culture* 51, no. 4 (2010): 893–918; Jennifer S. Light, "When Computers Were Women," *Technology and Culture* 40, no. 3 (1999): 455–483; and Ruth Oldenziel, *Making Technology Masculine: Men, Women and Modern Machines in America, 1870–1945* (Amsterdam: Amsterdam University Press, 1999). For works beyond history of technology explicitly related to the self, see also Donna Haraway, "A Cyborg Manifesto: Science, Technology, and Socialist-Feminism in the Late Twentieth Century," in *Simians, Cyborgs, and Women: The Reinvention of Nature* (New York: Routledge, 1990); Suchman, *Human–Machine Reconfigurations*; Sherry Turkle, *Life on the Screen: Identity in the Age of the Internet* (New York: Simon & Schuster, 1997); and N. Katherine Hayles, *How We Became Posthuman: Virtual Bodies in Cybernetics, Literature, and Informatics* (Chicago: University of Chicago Press, 1999).

36. Judy Wajcman, *Technofeminism* (New York: Polity, 2004), 107, emphasis added. Francesca Bray calls these technological practices "gynotechnics" and "androtechnics." See Bray, *Technology, Gender and History in Imperial China*, 17.

37. See Ruha Benjamin, *Race After Technology: Abolitionist Tools for the New Jim Code* (New York: Polity, 2019); Alondra Nelson, "Introduction: Future Texts," *Social Text* 20, no. 2 (2002): 1–15; Safiya Umoja Noble, *Algorithms of Oppression: How Search Engines Reinforce Racism* (New York: New York University Press, 2018). See also, Amy E. Slaton, *Race, Rigor, and Selectivity in U.S. Engineering: The History of an Occupational Color Line* (Cambridge, MA: Harvard University Press, 2010); Lundy Braun, *Breathing Race into the Machine: The Surprising Career of the Spirometer from Plantation to Genetics* (Minneapolis: University of Minnesota Press, 2014); Meredith Broussard, *Artificial Unintelligence: How Computers Misunderstand the World* (Cambridge, MA: MIT Press, 2018); Charlton D. McIlwain, *Black Software: The Internet and Racial Justice, from AfroNet to Black Lives Matter* (New York: Oxford University Press, 2019); André Brock Jr., *Distributed Blackness: African American Cybercultures* (New York: New York University Press, 2020); and Wendy Hui Kyong Chun, "Introduction: Race and/as Technology; or, How to Do Things to Race," *Camera Obscura* 24, no. 1 (2009): 7–35. On the construction of whiteness through technologies, see the essays in Bruce Sinclair, ed., *Technology and the African-American Experience* (Cambridge, MA: MIT Press, 2004); Evelyn Hammonds, "New Technologies of Race," in *Processed Lives: Gender and Technology in Everyday Life*, ed. Melodie Calvert and Jennifer Terry (New York: Routledge, 1997); David E. Nye, *America as Second Creation: Technology and Narratives of New Beginnings* (Cambridge, MA: MIT Press, 2003); Carolyn de la Peña, "The History of Technology, the Resistance of Archives, and the Whiteness of Race," *Technology and Culture* 51,

no. 4 (2010): 919–937; Gabrielle Hecht, *Residual Governance: How South Africa Foretells Planetary Futures* (Durham, NC: Duke University Press, 2023); and Tiago Saraiva, *The Orchard in Ruins* (Chicago: University of Chicago Press, forthcoming).

38. Sara Ahmed, *What's the Use?: On the Uses of Use* (Durham, NC: Duke University Press Books, 2019), 32.

39. See Arjun Appadurai and Neta Alexander, *Failure* (New York: Polity, 2019). For an overview of the growing interdisciplinary field of "failure studies," which includes breakdowns, see Neta Alexander, "Rage Against the Machine: Buffering, Noise, and Perpetual Anxiety in the Age of Connected Viewing," *Cinema Journal* 56, no. 2 (2017): 1-24. For an excellent discussion of the importance of failure for technology studies, see Larkin, *Signal and Noise: Media, Infrastructure, and Urban Culture in Nigeria*. On the queering of failure, see Jack Halberstam, *The Queer Art of Failure* (Durham, NC: Duke University Press, 2011).

40. On the historical role of technologies in creating difference, see, e.g., Wajcman, *Technofeminism* (New York: Polity, 2004); Bray, *Technology, Gender and History in Imperial China*; Lerman, "Categories of Difference, Categories of Power"; Oldenziel, *Making Technology Masculine; Nye, Technology Matters*, chap. 5; de la Peña, "The History of Technology, the Resistance of Archives, and the Whiteness of Race"; Slaton, *Race, Rigor, and Selectivity in U.S. Engineering*; and Tiago Saraiva, *The Orchard in Ruins*.

41. Ann Laura Stoler, *Race and the Education of Desire: Foucault's History of Sexuality and the Colonial Order of Things* (Durham, NC: Duke University Press, 1995), 7.

42. See Alfred Sohn-Rethel, *Das Ideal des Kaputten. Über neapolitanische Technik* (Bremen: Wassmann, 1926, repr. 1990), 33–38.

43. See, e.g., Michael Adas, *Machines as the Measure of Men: Science, Technology, and Ideologies of Western Dominance* (Ithaca, NY: Cornell University Press, 2015). See also aDaniel R. Headrick, *The Tools of Empire: Technology and European Imperialism in the Nineteenth Century* (New York: Oxford University Press, 1981). On the similar attempts to mark off the non-Western from the modern, see Dipesh Chakrabarty, *Provincializing Europe: Postcolonial Thought and Historical Difference* (Princeton, NJ: Princeton University Press, 2000), 77–83.

44. See, e.g., Schaffer, "Easily Cracked," 713. For a parallel concern over Western bodies, see Warwick Anderson, *The Cultivation of Whiteness: Science, Health and Racial Destiny in Australia* (Carlton, Australia: Melbourne University Press, 2002), 1.

45. See, e.g., Larkin, *Signal and Noise*, 246–247.

46. Larkin, *Signal and Noise*, 62.

47. See, e.g., Rudolf Mrázek, *Engineers of Happy Land: Technology and Nationalism in a Colony* (Princeton, NJ: Princeton University Press, 2002).

48. Achille Mbembe, *On the Postcolony* (Berkeley: University of California Press, 2019), 2.

49. On the revival of Heidegger in philosophy, science and technology studies, and postcolonial studies, see John Tresch, "Technological World-Pictures: Cosmic Things and Cosmograms," *Isis* 98, no. 1 (2007): 85.

50. For a recent evaluation of Heidegger's relationship to National Socialism in light of the publication of the *Black Notebooks*, see Richard Wolin, *Heidegger in Ruins: Between Philosophy and Ideology* (New Haven, CT: Yale University Press, 2023).

51. On "things" in history of science and technology, see, e.g., the papers introduced by Ken Alder, "Introduction," *Isis* 98, no. 1 (2007): 82. See also Lorraine Daston, ed., *Things That Talk. Object Lessons from the History of Art and Science* (New York: Zone Books, 2004). For an early and influential discussion of "thing theory," see Bill Brown, "Thing Theory," *Critical Inquiry* 28 (2001): 1–22.

52. Tiago Saraiva, *Fascist Pigs: Technoscientific Organisms and the History of Fascism* (Cambridge, MA: MIT Press, 2016), 237–240.

53. See, e.g., Galison, "On the Building, Crashing, and Thinking of Technologies and Selfhood."; Galison, "An Accident of History"; Star, "The Ethnography of Infrastructure"; Latour, *Reassembling the Social*; Verbeek, *What Things Do*, 79; Graham and Thrift, "Out of Order: Understanding Repair and Maintenance."; Jackson, "Rethinking Repair"; and Alexander, "Rage Against the Machine." *Cinema Journal* 56, no. 2 (2017): 14. For critique of this view, see Burrell, *Invisible Users*, 14.

54. The original German is "Dieses Unzuhandene stört und macht die Aufsässigkeit des zunächst und zuvor zu Besorgenden sichtbar." Martin Heidegger, *Sein Und Seit* (Tübingen, Germany: Max Niemeyer Verlag, 1967), 74. Heidegger takes up the "disturbing thing" in his later writings on technology in the postwar period. See Martin Heidegger, *The Question Concerning Technology, and Other Essays* (New York: Harper Torchbooks, 1977).

55. Oxford English Dictionary, "Disturbance."

56. See, e.g., Seth Reice, *The Silver Lining: The Benefits of Natural Disasters* (Princeton, NJ: Princeton University Press, 2001); Steward T. Pickett and P. S. White, *The Ecology of Natural Disturbance and Patch Dynamics* (San Diego, CA: Academic Press, 1985), 6–9; H. H. Lamb, *Climate, History, and the Modern World*, 2nd ed. (New York: Routledge, 1995); and Sander Govaerts, *Armies and Ecosystems in Premodern Europe* (Leeds: Arc Humanities Press, 2021), 105.

57. Anna Lowenhaupt Tsing, *The Mushroom at the End of the World* (Princeton, NJ: Princeton University Press, 2015), 5, 152. Incredibly, these two meanings—the psychological and the ecological—met in early twentieth-century scientific theories. On the connection between disturbance in early twentieth-century British ecological

theory and Freudian psychology, see Laura Cameron, "Histories of Disturbance," *Radical History Review*, no. 74 (1999): 5–24.

58. Tsing cites a similar interest in the dynamic relation between large-scale historical disturbances and the stable ecosystems of the open peasant forest. See Tsing, *The Mushroom at the End of the World*, 187.

59. Jones-Imhotep, *The Unreliable Nation*, chap. 1.

60. For the classic definition of technopolitics, see Gabrielle Hecht, *The Radiance of France: Nuclear Power and National Identity After World War II* (Cambridge, MA: MIT Press, 2009), 15. For an alternative usage, see Timothy Mitchell, *Rule of Experts: Egypt, Techno-Politics, Modernity* (Berkeley: University of California Press, 2002).

61. Jeffrey T. Schnapp, "The Romance of Caffeine and Aluminum," *Critical Inquiry* 28, no. 1 (2001): 245.

62. See, e.g., Edgerton, *The Shock of the Old*.

63. For a similar phenomenon in relation to scientific selves, see Daston and Galison, *Objectivity*.

64. On the similarly modern tendency to forget disaster, see Deborah R. Coen, "Conclusion," in *The Earthquake Observers: Disaster Science from Lisbon to Richter* (Chicago: University of Chicago Press, 2012).

65. The term "Atlantic World" is often meant to emphasize a set of material connections and circulations that I don't mean to invoke here. Although these various contexts are connected, I do not trace the intricate links that others do in histories of slavery, commerce, or even political thought. See, e.g., C. A. Bayly, *The Birth of the Modern World, 1780–1914* (London: Wiley-Blackwell, 2003); and Paul Gilroy, *The Black Atlantic: Modernity and Double Consciousness* (Cambridge, MA: Harvard University Press, 1993).

66. Marshall Berman describes these as the second and third phases of modernity. See Berman, *All That Is Solid Melts into Air*. My periodization coincides roughly with the emergence of the bourgeois order in the early nineteenth century, which Foucault saw as the beginning of the Modern Age. See Michel Foucault, *The Order of Things: An Archaeology of the Human Sciences* (New York: Routledge, 2002), xxii.

67. Tina M. Campt, *Listening to Images* (Durham, NC: Duke University Press, 2017), 5.

68. I have avoided the consistent use of the term "failure" here because of the confusion it creates. See Jones-Imhotep, *The Unreliable Nation*, 10–13.

69. On scientific controversies as moments of insight, see, e.g., Shapin and Schaffer, *Leviathan and the Air-Pump*, "Introduction."

70. For a similar approach to the history of liberalism, see Elaine Hadley, *Living Liberalism: Practical Citizenship in Mid-Victorian Britain* (Chicago: University of Chicago

Press, 2010). For a critique, see James Vernon, "What Was Liberalism, and Who Was Its Subject?; or, Will the Real Liberal Subject Please Stand Up?," *Victorian Studies* 53, no. 2 (2011): 303–310.

Chapter 1

1. On the flirtation of September 12, 1784, see Nicolas Edme Restif de La Bretonne, *Mes Inscripcions*, ed. Paul Cottin (Paris: Librairie Plon, 1889), 68, n414. On the murder of the widow Jamay (or Zamet) on January 31, 1790, see "Variétés," *Journal général de la cour et de la ville*, February 3, 1790, 4–5; and "Variétés," *Gazette nationale ou le Moniteur universel*, February 4, 1790, 141. For the establishments on Cour du Commerce, see Mathurin Roze de Chantoiseau, *Tablettes Royales de Renommée* (Paris: Royer, Bailly, Veuve Duchesne, Desnos and Buisson, 1786).

2. David A. Bell, "Little Spartas: What Causes Some Cities to Become Sites of Revolution?" *Nation*, January 25, 2018. On Desmoulins's residence in Cour du Commerce, see Jules Claretie, *Camille Desmoulins and His Wife: Passages from the History of the Dantonists*, trans. Frances Cashel Hoey (London: Smith, Elder and Co., 1876), 150–151.

3. Duplain was, in fact, a counterrevolutionary who formed part of the publishing network of the la Société Typographique de Neuchâtel (STN). On the STN and grub street publishers in eighteenth-century France, see Robert Darnton, *The Literary Underground of the Old Regime* (Cambridge, MA: Harvard University Press, 1982).

4. Hilary Mantel, "He Roared," *London Review of Books* 31, no. 15 (2009).

5. Tobias Schmidt, "Adresse au Comité d'Agriculture et de Commerce de Assemblée Nationale de France," January 3, 1791, Archives Nationales de France (ANF), F17 1136. Contemporary proposals for devices that allowed divers to work underwater seem to have been a common theme. See "Archives du Ministre de L'interieur, Mots de Recherche du Répertoire," no. 7, 1791, ANF, F17 1136.

6. Louis Du Bois, *Recherches historiques et physiologiques sur la guillotine, et détails sur Sanson* (Paris: France, 1843), 16. There is, however, no evidence placing Schmidt in Strasbourg in the winter and spring of 1792. See, e.g., Hector Fleischmann, *La Guillotine En 1793: D'après Des Documents inédits Des Archives Nationales* (Paris: Librairie des Publications Modernes, 1908), 47, n2.

7. See Achille Chereau, *Guillotin et la guillotine* (Paris: Bureaux de l'Union médicale, 1870), 26–27.

8. See, e.g., Kershaw, *A History of the Guillotine*; Arasse, *The Guillotine and the Terror*; Carol, *Physiologie de la veuve*; Outram, *The Body and the French Revolution*, chap.7; and Martschukat, "Ein Schneller Schnitt." On the French Revolution as a project of rationalization, see Ken Alder, *Engineering the Revolution: Arms and Enlightenment in France, 1763–1815* (Princeton, NJ: Princeton University Press, 1997). On the

complex meanings of "rationalization" during the period, see Ferenc Fehér, *The French Revolution and the Birth of Modernity* (Berkeley: University of California Press, 1990), 177–185.

9. On the guillotine as "ensign," see Pierre Jean Georges Cabanis, "Note sur l'opinion de MM. Olsmer et Sömmering, et du citoyen Sue, touchant le supplice de la guillotine," in *Œuvres philosophiques* (Paris: Presses universitaires de France, 1954). On the emphasis on the Terror in histories of the device, see Arasse, *The Guillotine and the Terror*; Carol, *Physiologie de la veuve*; Eamonn Carrabine, "The guillotine: Shadow, spectacle and the terror," *Crime, Media, Culture* 20, no. 4 (2023): 349-369.

10. The concern over the spectating public inspired rival proposals for execution technologies, including an "asphyxiation booth" and a "strangulation machine." On these and the guillotine as solutions to this concern, see Paul Friedland, *Seeing Justice Done: The Age of Spectacular Capital Punishment in France* (Oxford: Oxford University Press, 2012), chap. 10.

11. On the "emotional regime" of the French Revolution, see William M. Reddy, *The Navigation of Feeling: A Framework for the History of Emotions* (New York: Cambridge University Press, 2001). On the "normative order" for emotions, see Reddy, *The Navigation of Feeling*, 124. Carolyn Stearns and Peter Stearns have proposed a related term, "emotionology," which they define as the more general attitude toward "basic emotions and their appropriate expression." See Peter N. Stearns and Carol Z. Stearns, "Emotionology: Clarifying the History of Emotions and Emotional Standards," *The American Historical Review* 90, no. 4 (1985): 813–836. I use Reddy's formulation here, since it more explicitly links emotions to normativity and governance. For general histories of emotion, see Susan J. Matt, "Current Emotion Research in History: Or, Doing History from the Inside Out," *Emotion Review* 3, no. 1 (2011): 117–124; Nicole Eustace et al., "AHR Conversation: The Historical Study of Emotions," *American Historical Review* 117, no. 5 (2012): 1487–1531; and Rob Boddice and Mark Smith, *Emotion, Sense, Experience* (New York: Cambridge University Press, 2020).

12. See Arasse, *The Guillotine and the Terror*, 21. On the dangers of seeing and feeling too intensely, see Anne C. Vila, "Introduction: Powers, Pleasures, and Perils of the Senses in the Enlightenment Era," in *A Cultural History of the Senses in the Age of Enlightenment*, ed. Anne C. Vila (New York: Bloomsbury Academic, 2014), 7.

13. On technologies as "theaters of proof," see Noortje Marres and Javier Lezaun, "Materials and Devices of the Public: An Introduction," *Economy and Society* 40, no. 4 (2011): 489–509.

14. For histories of French sentimentalism, see footnote 17.

15. For a fuller treatment of this union of the mechanical arts and sentimental philosophy, see Edward Jones-Imhotep, "The Unfailing Machine: Mechanical Arts, Sentimental Publics and the Guillotine in Revolutionary France," *History of the*

Human Sciences 30, no. 4 (2017): 11–31. My use of the "sentimental machine" is inspired by John Tresch's analysis of "romantic machines." See John Tresch, *The Romantic Machine: Utopian Science and Technology After Napoleon* (Chicago: University of Chicago Press, 2012).

16. On the salons and the relation to the French public sphere, see Norbert Elias, *The Civilizing Process: The History of Manners*, vol. 1, trans. Edmund Jephcott (Oxford: Basil Blackwell, 1978), 39–40; Joan B. Landes, *Women and the Public Sphere in the Age of the French Revolution* (Ithaca, NY: Cornell University Press, 1988), 23–24; Jürgen Habermas, *The Structural Transformation of the Public Sphere: An Inquiry into a Category of Bourgeois Society*, trans. Thomas Burger (Cambridge, MA: MIT Press, 1989), 30; and Dena Goodman, *The Republic of Letters: A Cultural History of the French Enlightenment* (Ithaca, NY: Cornell University Press, 1994), 14.

17. For overviews of French sentimentalism, see Anne C. Vila, *Enlightenment and Pathology: Sensibility in the Literature and Medicine of Eighteenth-Century France* (Baltimore, MD: Johns Hopkins University Press, 1998), 225–257; William M. Reddy, "Sentimentalism and Its Erasure: The Role of Emotions in the Era of the French Revolution," *Journal of Modern History* 72, no. 1 (2000): 109–152; David Denby, *Sentimental Narrative and the Social Order in France, 1760–1820* (New York: Cambridge University Press, 1994); Jessica Riskin, *Science in the Age of Sensibility: The Sentimental Empiricists of the French Enlightenment* (Chicago: University of Chicago Press, 2002); Alex Hughes, "*The History of Tears: Sensibility and Sentimentality in France*. By Anne Vincent-Buffault," *French History* 6, no. 3 (1992): 381–382; Anne Vincent-Buffault, *The History of Tears: Sensibility and Sentimentality in France*, trans. Teresa Bridgeman (New York: St. Martin's Press, 1991); and Jan Goldstein, *The Post-Revolutionary Self: Politics and Psyche in France, 1750–1850* (Cambridge, MA: Harvard University Press, 2005). For a fascinating exploration of sentiment and technology during the Enlightenment, see Adelheid Voskuhl, *Androids in the Enlightenment: Mechanics, Artisans, and Cultures of the Self* (Chicago: University of Chicago Press, 2013).

18. Locke would also be a central figure in modern understandings of the self. See Charles Taylor, *Sources of the Self: The Making of the Modern Identity* (Cambridge, MA: Harvard University Press, 1989); and Seigel, *The Idea of the Self*.

19. Riskin, *Science in the Age of Sensibility*, 3. See also Vila, "Introduction: Powers, Pleasures, and Perils," 7.

20. Denby, *Sentimental Narrative and the Social Order in France*, 4.

21. Frank Baasner, *Der Begriff "sensibilité" im 18. Jahrhundert: Aufstieg und Niedergang eines Ideals* (Heidelberg: C. Winter, 1988), 135–136.

22. See Vincent-Buffault, *The History of Tears*, 9; Vila, *Enlightenment and Pathology*, 225–257; Denby, *Sentimental Narrative and the Social Order in France*, 139–140; and Riskin, *Science in the Age of Sensibility*, introduction.

23. Outram, *The Body and the French Revolution*; 122.

24. Rousseau and Cabanis are representative of these attitudes. See Martin S. Staum, *Cabanis: Enlightenment and Medical Philosophy in the French Revolution* (Princeton, NJ: Princeton University Press, 2014), 130. See also Elizabeth A. Williams, *The Physical and the Moral: Anthropology, Physiology, and Philosophical Medicine in France, 1750–1850* (New York: Cambridge University Press, 2002), 55; and Riskin, *Science in the Age of Sensibility*, 216.

25. Denby, *Sentimental Narrative and the Social Order in France*, 21.

26. Lynn Festa, *Sentimental Figures of Empire in Eighteenth-Century Britain and France* (Baltimore, MD: Johns Hopkins University Press, 2006), 3–5.

27. Riskin, *Science in the Age of Sensibility*, 214–215, 224–225. On the more general view of "the people" outside the Enlightenment, see Harvey Chisick, *The Limits of Reform in the Enlightenment: Attitudes Toward the Education of the Lower Classes in Eighteenth-Century France* (Princeton, NJ: Princeton University Press, 2014).

28. On sentimentalism and empire, see Festa, *Sentimental Figures*. On the racial contract, see Charles W. Mills, *The Racial Contract* (Ithaca, NY: Cornell University Press, 2014). The literature on Enlightenment Liberalism's tolerance for, and even support of, slavery is extensive. See, e.g., Louis Sala-Molins, *Le Code Noir, ou le Calvaire de Canaan* (Paris, 1987), 239–241; Wayne Glausser, "Three Approaches to Locke and the Slave Trade," *Journal of the History of Ideas* 51, no. 2 (1990): 199–216; Tyler Stovall, *White Freedom: The Racial History of an Idea* (Princeton, NJ: Princeton University Press, 2021), chaps. 3-4; Domenico Losurdo, *Liberalism: A Counter-History* (New York: Verso Books, 2014), chap. 2; Emmanuel Chukwudi Eze, *Race and the Enlightenment: A Reader* (Malden, MA: Blackwell, 1997); Peter Gay, *The Enlightenment: An Interpretation*, vol. 2, *The Science of Freedom* (New York: Knopf, 2013); Mary Efrosini Gregory, *Freedom in French Enlightenment Thought* (New York: Peter Lang, 2010); and Jonathan Israel, *A Revolution of the Mind: Radical Enlightenment and the Intellectual Origins of Modern Democracy* (Princeton, NJ: Princeton University Press, 2009).

29. See Festa, *Sentimental Figures*, chap. 3.

30. Vila, *Enlightenment and Pathology*; and Anne C. Vila, "Beyond Sympathy: Vapors, Melancholia, and the Pathologies of Sensibility in Tissot and Rousseau," *Yale French Studies* 92 (1997): 88–101.

31. Blindness formed a powerful object for sentimentalists' thinking about human knowledge and social development. See, e.g., Denis Diderot, "Lettre sur les aveugles a l'usage de ceux qui voyent" (1749) in *Ouvres philosophiques*, ed. Paul Vernière (Paris: Garnier, 1998). More broadly, see Riskin, *Science in the Age of Sensibility*, chap. 2.

32. Roze de Chantoiseau, *Tablettes Royales de Renommée*.

33. Vila, *Enlightenment and Pathology*, 210–224; Riskin, *Science in the Age of Sensibility*, 195–196; and Christiane Holm, "Sentimental Cuts: Eighteenth-Century Mourning Jewelry with Hair," *Eighteenth-Century Studies* 38, no. 1 (2004): 139–143.

34. On the vices of sentimentalism, see Reddy, *The Navigation of Feeling*, 161–172. On simplicity as a mechanical virtue emerging in the eighteenth century, see Agostino Ramelli, *The Various and Ingenious Machines of Agostino Ramelli (1588)*, trans. Martha Teach Gnudi (Baltimore, MD: Johns Hopkins University Press, 1976), 26.

35. Otto Mayr, *Authority, Liberty, and Automatic Machinery in Early Modern Europe* (Baltimore, MD: Johns Hopkins University Press, 1989), 124.

36. Gilbert Chinard, ed., *L'apothéose de Benjamin Franklin* (Paris: Librairie Orientale and Américaine, 1955), 27, cited in Riskin, *Science in the Age of Sensibility*, 136.

37. See, e.g., Louis Savot, *L'architecture françoise des bastimens particuliers* (Paris: François Clouvier and Pierre Aubouïn, 1673), 205. The French word for trouble here is *embrarras*.

38. See, e.g., "Prix préposés par la Société établie à Londres pour l'encouragement des Arts des Manufactures & du Commerce," *Gazette du Commerce*, August 1763, 181. On the balance, see Jean-François Trabaud, *Principes sur le mouvement et l'équilibre, pour servir d'introduction aux mécaniques et à la physique. Premier traité* (Paris: J. Desaint et C. Saillant, 1741), 430.

39. On the virtue of perfection, see Paola Bertucci, "Society of Arts," in *Artisanal Enlightenment: Science and the Mechanical Arts in Old Regime France* (New Haven, CT: Yale University Press, 2017), 53, 122.

40. Jacques de Vaucanson, "Construction de nouveaux moulins à organciner les foyes," *Mercure de France*, June 1751, 39.

41. See, e.g., Nicolas Grollier de Servière, *Recueil d'ouvrages curieux de mathématique et de mécanique, ou Description du cabinet de M. Grollier de Servière* (Lyon: D. Forey, 1719), 24.

42. Jean-Jacques Rousseau, "Lettres Ecrites de la montagne," in *Oeuvres complètes*, vol. 3 (Paris: Bénard et Compagnie, 1853), 14.

43. Claude Adrien Helvétius, "Letter to President Montesquieu," *Correspondance Générale D'Helvétius*, vol. 5, Appendicies et Index, ed. Alan Dainard et al. (Toronto: University of Toronto Press, 2004), 86.

44. "Nouvelles Etrangères," *Gazette du Commerce*, August 26, 1775, 537.

45. Félix Le Peletier de Saint-Fargeau, *Réflexions sur le moment présent, offertes à la Convention Nationale, par F. Lepelletier* (Paris: Imprimerie de R. Vatar, 1793), 9.

46. Honoré-Gabriel Riqueti de Mirabeau, *Travail sur l'éducation publique: Trouvé dans les papiers de Mirabeau l'ainé*, ed. P. J. G. Cabanis (Paris: Imprimerie Nationale, 1791), 82–83. These speeches were, in fact, drafted by Cabanis.

47. Honoré-Gabriel Riquieti de Mirabeau, *Mémoires biographiques, littéraires et politiques de Mirabeau écrits par lui-même, par son père, son oncle et son fils adoptif*, ed. Lucas Montigny (Paris: Delaunay Libraire, 1835), 7.147.

48. On the spectacles of the French Revolution, see Antoine De Baecque, *The Body Politic: Corporeal Metaphor in Revolutionary France, 1770–1800* (Stanford, CA: Stanford University Press, 1997), 251–252; and Mona Ozouf, *La Fête révolutionnaire 1789–1799* (Paris: Gallimard, 1976), 236–259. On the "omni-pedagogic environment," see Goldstein, *The Post-Revolutionary Self*, 84.

49. Jean-Jacques Rousseau, *Oeuvres complètes*, ed. Bernard Gagnebin and Marcel Raymond, vol. 4 (Paris: Gallimard, 1999), 517.

50. On the sentimental dangers of the crowd, see Fernando Vidal, *The Sciences of the Soul: The Early Modern Origins of Psychology*, trans. Saskia Brown (Chicago: University of Chicago Press, 2011), 2.

51. Cesare Beccaria, *Dei Delitti e Delle Pene* (Rome: Newton Compton, 2012); Thomas Paine, *Rights of Man, Common Sense, and Other Political Writings*, ed. Mark Philp (New York: Oxford University Press, 1998); and Foucault, *Discipline and Punish*. Paul Friedland argues that the Damiens execution crystallized shifting attitudes toward punishment in eighteenth-century France. See Friedland, *Seeing Justice Done*, chap. 7.

52. On the characteristics of the festivals, see Ozouf, *La Fête révolutionnaire 1789–1799*.

53. See Outram, *The Body and the French Revolution*, chap.7.

54. "Proposition, par M. Guillotin, de six articles relatifs aux suppliciés," October 10, 1789, Archives Parlementaires, C.S. 1, carton 33, dossier 303.

55. Denis Diderot and Pierre Tarin, "Anatomie," in *Encyclopédie, ou dictionnaire raisonné des sciences, des arts et des métiers, etc.*, ed. Denis Diderot and Jean le Rond d'Alembert, vol. 1 (Paris: Le Breton, 1751), 410.

56. Cited in Arasse, *The Guillotine and the Terror*, 3–4.

57. Étienne Charavay, *Revue des documents historiques: Suite de pièces curieuses et inédites, publiées avec des notes et des commentaires* (Paris: A. Lemerre, 1876), 58.

58. On the general concern over spectators at executions, see Friedland, *Seeing Justice Done*, chap. 10.

59. Foucault, *Discipline and Punish*, 91. See, e.g., Cesare Beccaria, *Traité des délits et des peines* (Paris: Hélie, 1856), 60, 87.

60. Pierre Louis de Lacretelle, *Discours sur le préjuge de peines infamantes* (Paris: Cuchet, 1784), 131.

61. Étienne Le Hodey de Saultchevreuil, ed., *Journal des États généraux* (Paris: Chez Devaux, 1789), 236–237. On the contemporary reluctance to speak about horrors, see Friedland, *Seeing Justice Done*, chap. 10.

62. Le Hodey de Saultchevreuil, *Journal des États généraux.*

63. *Rapport des commissaires chargés par le Roi, de l'examen du magnétisme animal* (Paris: Imprimerie Royale, 1784), 53–54.

64. On the image's tension between a public execution and a discreet one, see Friedland, *Seeing Justice Done*, 2.

65. On the theater of the guillotine, see Stephanie O'Rourke, "Beholder, Beheaded:: Theatrics of the Guillotine and the Spectacle of Rupture," in *Visual Culture and the Revolutionary and Napoleonic Wars*, ed. Satish Padiyar, Philip Shaw, and Philippa Simpson (New York: Routledge, 2016).

66. Arasse, *The Guillotine and the Terror*, 20; and G. Lenotre, *La Guillotine et les exécuteurs des arrêts criminels pendant la révolution, d'après des documents inédits tirés des Archives de L'état* (Paris: Perrin & Cie, 1920), 221.

67. See, e.g., Arasse, *The Guillotine and the Terror*, 186.

68. Letter from Verrier, Commissaire du roi, to Roederer, March 2, 1792, in Jules Antoine Taschereau, *Revue rétrospective, ou Bibliothèque historique*, ed. Jules Antoine Taschereau (Paris: L'Imprimerie de H. Fournier Ainé, 1835), 7.

69. Chereau, *Guillotin et la guillotine*, 12–16.

70. Carol, *Physiologie de la veuve*, 12.

71. William Smith, *Dictionary of Greek and Roman Antiquities* (New York: C. C. Little and J. Brown, 1853), 272.

72. On the category of the *artiste*, see Bertucci, *Artisanal Enlightenment*, 36, 133–134.

73. On the fascinating history of surgery in Ancien Régime France, see Charles C. Gillispie, *Science and Polity in France: The End of the Old Regime* (Princeton, NJ: Princeton University Press, 2004).

74. "Extrait de registres de l'academie chirurgie, du jeudi 31 décembre 1778," *L'Esprit des Journaux* 8, no 2 (1779): 344–345. Louis's official verification was important to guard against charlatanry (a topic that Louis himself associated with his additional battles against surgeon imposters). See Antoine Louis, "Sur les charlatans," Antoine Louis Papers, Bibliothèque de Metz, Réserve Précieuse MS 1310.

75. E. T. B. Saint-Edme, "Guillotine," in *Dictionnaire de la pénalité dans toutes les parties du monde connu* (Paris: Rousselon, 1828), 159–165.

76. E. T. B. Saint-Edme, "Guillotine," in *Dictionnaire de la pénalité dans toutes les parties du monde connu* (Paris: Rousselon, 1828), 159–165.

77. See Ken Alder, "Making Things the Same: Representation, Tolerance and the End of the Ancien Régime in France," *Social Studies of Science* 28, no. 4 (1998): 499–545.

78. "Abriss der auf dem Caroussel Plaze in Paris auf gebauten Bühne voohlange-brachte Köpf Maschine," Bibliothèque nationale de France, département Estampes et photographie, RESERVE FOL-QB-201.

79. Charavay, *Revue des documents historiques*, 48 (emphasis added).

80. This created the irony that the same mechanism used to drive the piles of the Louis XVI bridge would be used two years later and two hundred meters away to execute the King, after whom the bridge was named.

81. On the building boom, see Allan Potofsky, "The Construction of Paris and the Crises of the Ancien Régime: The Police and the People of the Parisian Building Sites, 1750–1789," *French Historical Studies* 27, no. 1 (2004): 9–48.

82. "Arts," *Journal de Paris*, February 8, 1787 (no. 39).

83. Alder, *Engineering the Revolution*, 87; and Raymonde Monnier, *Le faubourg Saint-Antoine, 1789–1815* (Paris: Bibliothèque d'histoire révolutionnaire, Société des études robespierristes, 1981), 66.

84. Alain Mercier, "Portefeuille Industriel: L'engrenage De La Prudence," *La revue* 18 (1997): 45–52.

85. The device's role in minting was so important that a bylaw, signed by Napoleon Bonaparte as First Consul in 1800, would decree that only manufacturers, goldsmiths, and clockmakers could possess one. They also had to keep it under lock and key when not in use. The original ordinance was issued in May 1796: "Ordonnance concernant la fabrication et l'emploi des Luminoirs, Moutons, Presses, Balanciers et Coupoirs," 4 Prairial an [23 mai 1796]. Napoleon signed a subsequent bylaw in 1800: "Arrêté relatif aux permissions nécessaires pour l'établissement de presses, moutons, laminoirs, balanciers et coupoirs." Archives Musée National des Techniques, H, N 184.

86. Chereau, *Guillotin et la guillotine*, 16–17.

87. Letter from Roederer to Guillotin cited in Letter from Verrier, Commissaire du roi, to Roederer, March 2, 1792, in Taschereau, *Revue rétrospective*, 8.

88. Charavay, *Revue des documents historiques*, 48–49.

89. One possibility was through Schmidt's diving machine, which had been inspected in the summer of 1790 by Antoine Portal, member of the Académie des Sciences and an expert, like Louis, on drownings. See Antoine Portal, "Avis Important sur les moyens pratiqués avec succès pour secourir" (Paris: L'Imprimerie Royale, 1776). Portal had also been called to the Cour du Commerce in 1782, before Schmidt arrived in Paris, to attend to the daughter of a carpenter, who had fallen ill there. Antoine Portal, *Observations sur la nature et sur le traitement de la phthisie pulmonaire* (Paris: DuPont, 1792), 124.

90. Canton de Paris, "Registre des actes de publication de mariage pour la municipalité du onzième arrondissement," entry no. 86, Archives de Paris, AD.75 Cote: Vbis6E1/1.

91. "Réception de maitres et mattresses de métiers," July 28, 1785, ANF, Fonds Châtelet Y 9395A.

92. *Calendrier Musical Universel* (Paris: Prault, 1788), 285.

93. *Moniteur*, octedi, 8 vendémiaire an III (29 septembre, 1794), cited in Chereau, *Guillotin et la guillotine*, 28.

94. "Arts," *Journal de Paris*, February 10, 1789 (no. 41): 179–180.

95. About $50,000 in current US dollars. Chereau, *Guillotin et la guillotine*, 19–20.

96. Charavay, *Revue des documents historiques*, 52.

97. On the "unthinkability" of the Haitian Revolution at the time, see Michel-Rolph Trouillot, *Silencing the Past: Power and the Production of History* (Boston, MA: Beacon Press, 2015), chap. 3.

98. Letter from Roederer to Provisional Second Tribunal, April 11, 1792, in Taschereau, *Revue rétrospective*, 14–15; and Letter from Roederer to Clavière, June 7, 1792, in Taschereau, *Revue rétrospective*, 24.

99. Letter from Louis to Roederer, April 19, 1792, in Charavay, *Revue des documents historiques*, 53.

100. Letter from Roederer to Challan, April 19, 1792, in Taschereau, *Revue rétrospective*, 16, 18.

101. See Jeff Horn, "The Other 'Great Fear': Labor Relations, Industrialization, and Revolution," in *The Path Not Taken: French Industrialization in the Age of Revolution, 1750–1830* (Cambridge, MA: MIT Press, 2008).

102. Charavay, *Revue des documents historiques*, 54.

103. *Chronique de Paris*, April 16, 1792 (no. 118).

104. Chereau, *Guillotin et la guillotine*, 23. On the earlier enthusiasm for public executions and the disappointments of the guillotine, see Friedland, *Seeing Justice Done*, chap. 7–10.

105. J. de la Tynna, *Almanach du commerce de Paris, des départemens de la France et des principales villes du monde Année 1811* (Paris: de la Tynna, 1811), 206.

106. Chereau, *Guillotin et la guillotine*, 28.

107. Carol, *Physiologie de la veuve*, 56.

108. Giraud, "Rapport sur la machine à décapiter," June 5, 1792, in Taschereau, *Revue rétrospective*, 22–23.

109. Marc-Antoine Laugier quoted in Potofsky, "The Construction of Paris," 31.

110. Letter from Roederer to the Minister of Public Contributions, June 7, 1792, in Taschereau, *Revue rétrospective*, 24–26.

111. The Ministers of Public Contributions were falling fast. On 19 July, Beaulieu was replaced by René Delaville-Le Roulx, who wrote to Roederer on August 6, four days before he himself would be replaced.

112. Letter from Roederer to the Minister of Public Contributions, July 13, 1792, in Taschereau, *Revue rétrospective*, 28.

113. Clairin is listed as *ébeniste*, an inlayer, among the many furniture and cabinet-makers on Cour du Commerce in 1780. Henri Clouzot, *Les Meubles du VIIIe siècle* (Paris: Albert Monacé, 1780), 239. He is still there in 1798. See de la Tynna, *Almanach du Commerce de Paris*, 320.

114. Letter from Roederer to the Minister of Public Contributions, July 13, 1792, in Taschereau, *Revue rétrospective*, 29.

115. Letter from Roederer to the Minister of Public Contributions, July 17, 1792, in Taschereau, *Revue rétrospective*, 29–31.

116. Letter from Roederer to the Minister of Public Contributions, July 17, 1792, in Taschereau, *Revue rétrospective*, 29–31.

117. "Réponse du sieur Schmidt aux faits contenus dans la lettre de M. le procureur géneral syndic du département de Paris, du 28 juillet 1792, l'an IV de la liberté, qui a été communiqué par ordre du ministre, relativement à la machine à décapiter," in Chereau, *Guillotin et la guillotine*, 30–32.

118. Carol, *Physiologie de la veuve*, 56–57.

119. Letter from the Minister of Justice to the Minister of the Interior, March 26, 1796, ANF, F17 1136.

120. "Exposition des produits de l'Industrie Française à Paris, 1806," Archives Musée National des Techniques, P293, no. 194.

Chapter 2

1. Nicholas Daly, "Railway Novels: Sensation Fiction and the Modernization of the Senses," *ELH* 66, no. 2 (1999): 461-462.

2. Schivelbusch, *Railway Journey*, 131.

3. On the railway as an early and emblematic large-scale system, see, e.g., Leo Marx, "Technology: The Emergence of a Hazardous Concept," *Social Research* 64, no. 3 (1997): 972. On the contemporary Victorian understanding of the railway "system" see Ben Marsden and Crosbie Smith, *Engineering Empires: A Cultural History of Technology in Nineteenth-Century Britain* (London: Palgrave Macmillan UK, 2005), 145-150. On the devastating dynamics of systems, see Perrow, *Normal Accidents*.

4. On large technological systems, see Hughes, *Networks of Power*; Agatha C. Hughes and Thomas P. Hughes, eds., *Systems, Experts, and Computers: The Systems Approach in Management and Engineering, World War II and After* (Cambridge, MA: MIT Press, 2000); Michael Thad Allen and Gabrielle Hecht, eds., *Technologies of Power: Essays in Honor of Thomas Parke Hughes and Agatha Chipley Hughes* (Cambridge, MA: MIT Press, 2001); Marx, "Technology: The Emergence of a Hazardous Concept"; Bijker et al., *The Social Construction of Technological Systems*; and Richard Rhodes, ed., *Visions of Technology: A Century of Vital Debate about Machines, System, and the Human World* (New York: Simon and Schuster, 1999). On the "normal accidents" these systems produce, see Perrow, *Normal Accidents*.

5. Tom Crook, *Governing Systems: Modernity and the Making of Public Health in England, 1830–1910* (Berkeley: University of California Press, 2016), 4.

6. "Report of the Offices of the Railway Department, 1844–1845," *Parliamentary Papers* 39 (London: HM Stationery Office, January 1846), 55–63.

7. For the most comprehensive overview of the history of the Royal Corps of Engineers up to the late 19th century, see Whitworth Porter, *History of the Corps of Royal Engineers* (London: Longmans, Green, 1889). On the place of engineering in British Imperialism, see Iwan Rhys Morus, *How the Victorians Took Us to the Moon: The Story of the 19th-Century Innovators Who Forged Our Future* (New York: Pegasus Books, 2022), 17-22. On the importance of the Ordinance Survey for the British imperial project, see Vernon, *Distant Strangers*, 57-58. On the broader role of engineers in nineteenth-century liberal governance, see Michel Foucault, "Space, Knowledge, and Power," in Paul Rabinow (ed.) *The Foucault Reader* (London: Penguin, 1986), 244; and Joyce, *Rule of Freedom*, chap. 2.

8. Vernon, *Distant Strangers*, 64. On the new role of engineers in Victorian administration, see Joyce, *Rule of Freedom*, chap. 2.

9. Janet Cutler, "Neglected Historical Sources: The Railway Inspectors Reports." *History Workshop Journal* 27, no. 1 (1989): 175; and Henry Parris, *Government and the Railways in Nineteenth-Century Britain* (London: Routledge & K. Paul, 1965), 32.

10. Between 1840 and 1891, fifteen inspectors served this function in total. See Jack Simmons, *Express Train and Other Railway Studies* (Nairn: D. St. J. Thomas, 1994), 214.

11. On this larger construction of a safe England, see Elaine Freedgood, *Victorian Writing About Risk: Imagining a Safe England in a Dangerous World* (New York: Cambridge University Press, 2006), 1.

12. Iwan Rhys Morus, *How the Victorians Took Us to the Moon: The Story of the 19th-Century Innovators Who Forged Our Future* (New York: Pegasus Books, 2022), 85.

13. Samuel Smiles, *Self-help, with Illustrations of Character and Conduct* (London: John Murray, 1859), 2.

14. For the prominent place of engineers and inventors in Smiles's thought and writing, see Morus, *How the Victorians Took Us to the Moon*, 85-86. For the more general cultural role of the Victorian railway engineer, see Marsden and Smith, *Engineering Empires*, chap. 4.

15. Quoted in Schivelbusch, *Railway Journey*, 131–132 (emphasis added).

16. Michael Reynolds, *Engine-Driving Life* (London: Crosby Lockwood and Co., 1881), 25, 54–55.

17. See, e.g., Henry W. Disney, *The Law of Carriage by Railway* (London: Stevens and Sons, 1905), 129, cited in R. W. Kostal, *Law and English Railway Capitalism, 1825–1875* (Oxford: Clarendon, 1994), 256. See also Ralph Harrington, "The Railway Accident: Trains, Trauma, and Technological Crisis in Nineteenth-Century Britain," in *Traumatic Pasts: History, Psychiatry, and Trauma in the Modern Age, 1870–1930*, ed. Mark S. Micale and Paul Lerner (Cambridge: Cambridge University Press, 2010), 33.

18. James Taylor, "Business in Pictures: Representations of Railway Enterprise in the Satirical Press in Britain 1845–1870," *Past and Present* 189, no. 1 (2005): 113.

19. Simmons, *Express Train and Other Railway Studies*, 219; and Parris, *Government and the Railways*, 47-8. See also Cutler, "Neglected Historical Sources," 175.

20. Public Generals Acts, 3&4 Vict., cap. 97, sec. 3, 5.

21. Later returns would require the name of the injured, the nature of the injury, and any "other details." On the role of standardized forms, trained personnel, and machines in Victorian state power, see Vernon, *Distant Strangers*, 62.

22. Schivelbusch, *Railway Journey*, xiv.

23. On the earlier understanding of accidents as events with no cause, see Cooter and Luckin, "Accidents in History"; Tom Crook and Mike Esbester, "Risk and the History of Governing Modern Britain, c. 1800–2000," in *Governing Risks in Modern Britain: Danger, Safety and Accidents, c. 1800–2000*, ed. Tom Crook and Mike Esbester (London: Palgrave Macmillan, 2016), 21; and Samuel Johnson, "Accident," in *A Dictionary of the English Language* (London: W. Strahan, 1755).

24. See Cutler, "Neglected Historical Sources."

25. During the first decade of the nineteenth century, *The Times* coverage of single disasters rose 90 percent. See Ryan Vieira, "Rethinking the History of the Risk Society: Accident Reporting, the Social Order and the London Daily Press, 1800–30," in *Governing Risks in Modern Britain: Danger, Safety and Accidents, c. 1800–2000*, ed. Tom Crook and Mike Esbester (London: Palgrave Macmillan, 2016), 60.

26. Vieira, "Rethinking the History of the Risk Society," 57. On the blamelessness of narratives, see Bronstein, *Caught in the Machinery*, quoted in Vieira "Rethinking the History of the Risk Society," 61, n55. By the mid-nineteenth century, Bronstein

claims that this form of narrative—focused on the single accident, avoiding questions of cause, and centered on humans—had become a central way for Victorians to make sense of industrial modernity. See Vieira, "Rethinking the History of the Risk Society," 75.

27. For a description of the procedures surrounding the reports, and their eventual impact on public opinion, see Simmons, *Express Train and Other Railway Studies*, 217–229.

28. See, e.g., Thomas P. Hughes, "The Evolution of Large Technological Systems," in *The Social Construction of Technological Systems*, ed. Wiebe E. Bijker et al. (Cambridge, MA: MIT Press, 2012), 45–76.

29. The Oxford English Dictionary cites the 1734 usage by Desaguliers.

30. Samuel Johnson, *A Dictionary of the English Language*, s.v. "system (n.s.), sense 3," (London: W. Strahan, 1755). See also Crook, *Governing Systems*, 4, n8.

31. Vernon, *Distant Strangers*, 52, 70–72. Tom Crook argues that modernity in Victorian Britain lies in the making and remaking of systems rather than in the making and remaking of the state. See Crook, *Governing Systems*, 11. On Victorian Britain's "society of strangers" in relation abstract systems, see Vernon, *Distant Strangers*. On the new role of engineers in Victorian administration, see Joyce, *Rule of Freedom*, chap. 2.

32. Crook, *Governing Systems*, 148.

33. See Patrick Joyce, *The State of Freedom: A Social History of the British State Since 1800* (Cambridge: Cambridge University Press, 2013); and Mitchell, *Rule of Experts*.

34. See Joyce, *The Rule of Freedom*.

35. Crook, *Governing Systems*, 11; and Vernon, *Distant Strangers*, 14.

36. Vernon, *Distant Strangers*, 15. For the case of public health, see Crook, *Governing Systems*, 288.

37. "Report of the Offices of the Railway Department, 1844–1845," 140–142.

38. "Report of the Offices of the Railway Department, 1844–1845," 142.

39. Timothy L. Alborn, *Conceiving Companies: Joint-Stock Politics in Victorian England* (New York: Routledge, 1998); Frank Dobbin, *Forging Industrial Policy: The United States, Britain, and France in the Railway Age* (Cambridge: Cambridge University Press, 1994); and Kostal, *Law and English Railway Capitalism*. See also Taylor, "Business in Pictures."

40. "Report to Committee of Privy Council for Trade on Accidents on Railways, 1855," *Parliamentary Papers* 54 (London: HM Stationery Office, January 1856), 14.

41. Harrington, "The Railway Accident," 33–34.

42. On the railway and British identity, see, e.g., Vernon, *Distant Strangers*; Adas, *Machines as the Measure of Men*; and Michael J. Freeman, *Railways and the Victorian Imagination* (New Haven, CT: Yale University Press, 1999).

43. Vieira, "Rethinking the History of the Risk Society," 61, 73.

44. See Ralph Harrington, "Railway Safety and Railway Slaughter: Railway Accidents, Government and Public in Victorian Britain," *Journal of Victorian Culture* 8 (2003): 187–207.

45. Tina Young Choi, "Writing the Victorian City: Discourses of Risk, Connection, and Inevitability," *Victorian Studies, 43,* no. 4 (2001): 561–589.

46. On the specific connections to Dickens, see Jill L. Matus, "Trauma, Memory, and Railway Disaster: The Dickensian Connection," *Victorian Studies* 43, no. 3 (2001): 413–436.

47. Schivelbusch, *Railway Journey*, 131.

48. Kostal, *Law and English Railway Capitalism*, 281.

49. Schivelbusch, *Railway Journey*. See also Harrington, "The Railway Accident," 34; and Daly, "Railway Novels," 468.

50. On the common language of internal destruction mirrored between human and machine, see Matthew Wilson Smith, "Victorian Railway Accident and the Melodramatic Imagination." *Modern Drama* 55, no. 4 (2012): 504; and Schivelbusch, *Railway Journey*.

51. *The Influence of Railway Travelling on Public Health* (London: Hardwicke, 1862), 20.

52. On fears of lack of control, see Harrington, "The Railway Accident," 36.

53. *The Influence of Railway Travelling on Public Health*, 43.

54. "Reports of Inspecting Officers of Railway Dept. on Accidents on Railways, May–December 1865," *19th Century House of Commons Sessional Papers* Volume Page: LXIII.115 Volume: 63 Paper Number: 3609 (January 1, 1866), 42.

55. "The Inquest Upon the Suffers in the Late Accident," *Times* [London, England], 1865 (emphasis added).

56. Henry Whatley Tyler, "The Official Reports and Returns of the Railway Department of the Board of Trade," *Quarterly Review* 3, no. 221 (1862).

57. Tyler, "The Official Reports and Returns," 5.

58. Tyler, "The Official Reports and Returns," 36.

59. *The Lancet* identified signaling as a third major cause of accidents, after collisions from overwork and failures of wheels. See *The Influence of Railway Travelling on Public Health*, 132.

60. See, e.g., Douglas Galton, "Report to the Lords of the Committee of the Privy Council for Trade upon the Accidents Which Have Occurred on Railways During the Year 1858," C Series (London: HM Stationery Office, 1859), 17–18. See also reports for 1868, third part, page 40.

61. See, e.g., "Reports of the Inspecting Officers of the Railway Department," *Parliamentary Papers* 57 (London: HM Stationery Office, 1861), 137–138.

62. Douglas Galton, "Report to the Lords of the Committee of Privy Council for Trade and Foreign Plantations of the Proceedings of the Department Relating to Railways for the Year 1858," Appendix 1 (London: HM Stationery Office, 1856), 33.

63. John Saxby, "Railway and Signals," UK Patent No. 1479, 1856.

64. Précis of Official Papers, 1881, Part VII, page 39.

65. *Iron*, October 10, 1879.

66. On the general system of discipline for railway servants, see P. W. Kingsford, *Victorian Railwaymen* (London: Frank Cass and Co. Ltd., 1970), chap. 2.

67. *Oxford English Dictionary*, s.v. "demeanour | demeanor (n.), sense 1.a," December 2023.

68. See "Punishments and Awards, Enginemen: Register, South Eastern Railway Company (1869–1944)," RAIL 635/305, UK National Archives, Kew. For other examples of railway punishment books, see "Bishopsgate Station, Misdemeanor and Punishment book, 1892–1894," RAIL 227/342; and "Punishments for Serious Offences between March and December (1874), Great Northern Railway Company: Records," RAIL 236/316/12. UK National Archives, Kew.

69. "General Report to the Board of Trade upon the Accidents Which Have Occurred on the Railways of the United Kingdom During the Year 1881," Excerpt, August 24, 1881 (London: HM Stationery Office, 1882), 34 and 35.

70. "General Report to the Board of Trade," 34, 35.

71. Vernon, *Distant Strangers*, 16.

Chapter 3

1. In its engagement with archival methods and Black storytelling, this chapter is deeply indebted to the theoretical work of Saidiya Hartman, Hazel Carby, Tiya Miles, Jennifer Morgan, Marisa Fuentes, and Katherine McKittrick. For representative works, see Saidiya Hartman, *Wayward Lives, Beautiful Experiments: Intimate Histories of Riotous Black Girls, Troublesome Women, and Queer Radicals* (New York: Norton & Company, 2019); Hazel V. Carby, *Imperial Intimacies: A Tale of Two Islands* (New York: Verso , 2019); Tiya Miles, *All That She Carried: The Journey of Ashley's Sack, a Black Family Keepsake* (New York: Random House, 2021); Jennifer L. Morgan, *Reckoning with*

Slavery: Gender, Kinship, and Capitalism in the Early Black Atlantic (Durham: Duke University Press, 2021); Marisa J. Fuentes, *Dispossessed Lives: Enslaved Women, Violence, and the Archive* (Philadelphia: University of Pennsylvania Press, 2016); Katherine McKittrick, *Dear Science and Other Stories* (Durham, NC: Duke University Press, 2021).

2. George G. Foster, *New York By Gas-Light and Other Urban Sketches* (Berkeley: University of California Press, 1990), 145.

3. Charles Dickens, *American Notes for General Circulation* (London: Chapman and Hall, 1850), 62.

4. On the historical invisibility of Almack's and the social forces creating the Five Points, see James W. Cook, "Dancing Across the Color Line: A Story of Mixtures and Markets in New York's Five Points," *Commonplace* 4, no. 1 (2003).

5. Cook, "Dancing Across the Color Line."

6. Cook, "Dancing Across the Color Line."

7. Claire Perry, *Young America: Childhood in 19th-Century Art and Culture* (New Haven, CT: Yale University Press, 2006), 101–103. On hidden transcripts, see James C. Scott, *Domination and the Arts of Resistance: Hidden Transcripts* (New Haven, CT: Yale University Press, 1990); and Robin D. G. Kelley, "'We Are Not What We Seem': Rethinking Black Working-Class Opposition in the Jim Crow South," *Journal of American History* 80, no. 1 (1993): 75–112.

8. Thomas F. de Voe, *The Market Book, Containing a Historical Account of the Public Markets in the Cities of New York, Boston, Philadelphia and Brooklyn, with a Brief Description of Every Article of Human Food Sold Therein, the Introduction of Cattle in America, and Notices of Many Remarkable Specimens*, vol. 1 (New York: Printed for the author, 1862), 322, 344. On the Pinkster festival, see White, Shane. "'It Was a Proud Day': African Americans, Festivals, and Parades in the North, 1741-1834," *The Journal of American History* 81, no. 1 (1994): 13-50.

9. For the larger cultural theft that blackface performance represented, see Eric Lott, *Love and Theft: Blackface Minstrelsy and the American Working Class* (New York: Oxford University Press, 1993). See also Andrew Womack, "Ridicule and Wonder: The Beginnings of Minstrelsy and New York," *Afro-Americans in New York Life and History* 36, no. 2 (2012): 86–115.

10. On Juba and Almack's, see Lott, *Love and Theft*, 112-116.

11. "Great Public Contest," *The New York Herald* 10, no. 188 (July 8, 1844), quoted in Cook, "Dancing across the Color Line." For a remarkable reconstruction of Lane's life and death, see James W. Cook, "Lane, William Henry [*performing name* Juba, Master Juba] (c. 1824–c. 1854), Dancer," *Oxford Dictionary of National Biography* (October 3, 2013).

12. Cook, "Dancing across the Color Line."

13. Marian Hannah Winter, "Juba and American Minstrelsy," *Dance Index* 6, no. 2 (1947): 28–47.

14. On Congo Square in the history of jazz, see Ted Gioia, *The History of Jazz*, 3rd ed. (New York: Oxford University Press, 2021), 1–6. On the dance of the breakdown in Congo Square, see Benjamin M. Norman, *Norman's New Orleans and Environs* (New Orleans: B. M. Norman, 1845), 182; *New Orleans Daily Picayune*, June 24, 1845, quoted in Gary A. Donaldson, "A Window on Slave Culture: Dances at Congo Square in New Orleans, 1800–1862," *Journal of Negro History* 69, no. 2 (1984): 67, 72, n25; Henry A. Kmen, *Music in New Orleans: The Formative Years, 1791–1841* (Baton Rouge: Louisiana State University Press, 1966), 229; and Henry A. Kmen, "The Roots of Jazz and the Dance in Place Congo: A Re-Appraisal," *Anuario Interamericano de Investigacion Musical* 8 (1972): 11. In Haiti, where it was called "Djouba," the dance was performed by farmers and peasants, and celebrated the land and its fruits. On the Juba as a competitive dance of skill, see Joseph E. Holloway, *Africanisms in American Culture*, 2nd ed. (Bloomington: Indiana University Press, 2005), 52.

15. Charles B. Hersch, *Subversive Sounds: Race and the Birth of Jazz in New Orleans* (Chicago: University of Chicago Press, 2008), 4–5; and Joel Dinerstein, *Swinging the Machine: Modernity, Technology, and African American Culture Between the World Wars* (Amherst: University of Massachusetts Press, 2003).

16. Burke, "Bursting Boilers and the Federal Power," 18. See also Moon, *The Machines of Leonardo da Vinci and Franz Reuleaux*, 252.

17. Stephanie Camp, *Closer to Freedom: Enslaved Women and Everyday Resistance in the Plantation South* (Chapel Hill: University of North Carolina Press, 2004), xxix. The urban slavery of the Caribbean provided another context for rival geographies. See Fuentes, *Dispossessed Lives*. On the role of steamboats in creating this geography, see Thomas C. Buchanan, "Rascals on the Antebellum Mississippi: African American Steamboat Workers and the St. Louis Hanging of 1841," *Journal of Social History* 34, no. 4 (2001): 797–816.

18. For the most comprehensive study of this world, see Thomas C. Buchanan, *Black Life on the Mississippi: Slaves, Free Blacks, and the Western Steamboat World* (Chapel Hill: University of North Carolina Press, 2004).

19. Henry Bibb, "Narrative of the Life and Adventures of Henry Bibb, An American Slave," in *Puttin' on Ole Massa: The Slave Narratives of Henry Bibb, William Wells Brown, and Solomon Northop*, ed. Gilbert Osofsky (New York: Harper and Row, 1969), 84.

20. Brown later worked on a Lake Erie steamboat, helping slaves travel from Detroit and Buffalo into Canada after 1850. William Farmer, ed., "Memoir of William Wells Brown," in *Three Years in Europe: Or, Places I Have Seen and People I Have Met* (New York: Cambridge University Press, 2014), xviii.

21. Henry Bibb, *Narrative of the Life and Adventures of Henry Bibb, an American Slave, Written by Himself* (Chapel Hill: University of North Carolina at Chapel Hill Library,

2018), 49; and Bibb, "Narrative of the Life and Adventures of Henry Bibb, An American Slave," 84.

22. Buchanan, *Black Life on the Mississippi*, 171.

23. See Edward Jones-Imhotep and Alexander Offord, "Black Steam: Patents, Portals, and the Counter-Histories of the Victorian Android," in *Victorian Automata: Mechanism and Agency in the Nineteenth Century*, ed. Suzy Anger and Thomas Vranken (New York: Cambridge University Press, 2024).

24. Frederick Law Olmsted, *A Journey in the Seaboard Slave States, With Remarks on Their Economy* (New York: Mason Brothers, 1861), 426–427. Olmsted's own project of Central Park would displace the Black community of Seneca Village.

25. W. E. B. Du Bois, "The Negro Artisan," in *The Negro Artisan: Report of a Social Study Made under the Direction of Atlanta University; Together with the Proceedings of the Seventh Conference for the Study of the Negro Problems, Held at Atlanta University, on May 27th, 1902*, ed. W. E. B. Du Bois (Atlanta, GA: Atlanta University Press, 1902), 17–18.

26. For a history of rascality as an expression of Black freedom, see Buchanan, "Rascals on the Antebellum Mississippi." For clothing as a source of antebellum Black expression and a technology of control, see Shane White and Graham White, *Stylin': African-American Expressive Culture, from Its Beginnings to the Zoot Suit* (Ithaca, NY: Cornell University Press, 1999).

27. Buchanan, *Black Life on the Mississippi*, 125.

28. Buchanan, "Rascals on the Antebellum Mississippi," 802–803; and Buchanan, *Black Life on the Mississippi*, 129. Robin D. G. Kelley has explored similar practices of workplace resistance in the Jim Crow South. See Robin D. G. Kelley, *Race Rebels: Culture, Politics, and the Black Working Class* (New York: Free Press, 1996).

29. Buchanan, *Black Life on the Mississippi*, 128.

30. Samuel A. Cartwright, "Report on the Diseases and Physical Peculiarities of the Negro Race," *New Orleans Medical and Surgical Journal* VII (May 1851): 691–715. In the late eighteenth century, slaves performed most mechanical occupations in the Lower South. As Joyce Chaplin notes, claims that slaves could not be artisans or machine operators were inventions of antebellum proslavery campaigns. See Joyce Chaplin, *An Anxious Pursuit: Agricultural Innovation and Modernity in the Lower South, 1730–1815* (Chapel Hill: University of North Carolina Press, 1993), 353–354. On the presumed tendency of Black workers to break tools, or cause breakdowns in machinery, see Seth Rockman, *Plantation Goods: A Material History of American Slavery* (Chicago: University of Chicago Press, 2024), 211.

31. Cartwright, "Report on the Diseases and Physical Peculiarities of the Negro Race." For similar pathologizations of slave escape and resistance in Haiti, Jamaica, and the Guianas, see Trouillot, *Silencing the Past*, 83–84.

32. Judith A. Carney, *Black Rice: The African Origins of Rice Cultivation in the Americas* (Cambridge, MA: Harvard University Press, 2009).

33. Sven Beckert, "Slavery Takes Command," in *Empire of Cotton: A Global History* (New York: Knopf, 2014).

34. On the role of rice cultivation in this negotiation, see Carney, *Black Rice*. On the place of slave resistance and rebellion in shaping the antebellum South, see, e.g., Herbert Shapiro, "The Impact of the Aptheker Thesis: A Retrospective View of *American Negro Slave Revolts*," *Science and Society* 48, no. 1 (1984): 52–73.

35. An example is James Henry Hammond, who was unable to force slaves at the edge of the Low Country to shift from task-based to gang-based methods of cotton cultivation in the 1830s. See Drew Gilpin Faust, *James Henry Hammond and the Old South: A Design for Mastery* (Baton Rouge: Louisiana State University Press, 1985), 69–104. On the task system, see also Chaplin, *An Anxious Pursuit*, 326–327; Charles Joyner, *Down by the Riverside: A South Carolina Slave Community* (Urbana: University of Illinois Press, 1984), 43–46; Caitlin Rosenthal, *Accounting for Slavery: Masters and Management* (Cambridge, MA: Harvard University Press, 2018), 85–120; Edward E. Baptist, "Toward a Political Economy of Slave Labor: Hands, Whipping-Machines, and Modern Power," in *Slavery's Capitalism*, ed. Sven Beckert and Seth Rockman (Philadelphia: University of Pennsylvania Press, 2016), 38; Peter A. Coclanis, "How the Low Country Was Taken to Task: Slave-Labor Organization in Coastal South Carolina and Georgia," in *Slavery, Secession, and Southern History*, ed. Robert Louis Paquette and Louis A. Ferleger (Charlottesville: University of Virginia Press, 2000); Philip D. Morgan, "Task and Gang Systems: The Organization of Labor on New World Plantations," in *Work and Labor in Early America*, ed. Stephen Innes (Chapel Hill: University of North Carolina Press, 1988); and Philip D. Morgan, *Slave Counterpoint: Black Culture in the Eighteenth-Century Chesapeake and Lowcountry* (Chapel Hill: University of North Carolina Press, 1998), 179–194. The task system did nothing to protect slaves or their children from the health risks of the swampy Low Country environment, and they suffered and died in vast numbers from malaria throughout the years of slavery. See William Dusinberre, *Them Dark Days: Slavery in the American Rice Swamps* (New York: Oxford University Press, 1996).

36. James Scott has called these techniques of "everyday resistance." See James C. Scott, "Everyday Forms of Resistance," in *Everyday Forms of Peasant Resistance*, ed. Forrest D. Colburn (New York: Routledge, 2015), 3–33. Building on Scott's notion of infrapolitics, Robin D. G. Kelley argues that these practices are diagnostic of power. See Kelley, "'We Are Not What We Seem.'"

37. Faust, *James Henry Hammond and the Old South*, 92–94.

38. Baptist, "Toward a Political Economy of Slave Labor," 35.

39. Angela Lakwete, *Inventing the Cotton Gin: Machine and Myth in Antebellum America* (Baltimore, MD: Johns Hopkins University Press, 2005), 31.

40. See Chaplin, *An Anxious Pursuit*, 353–354. See also Bruce Sinclair, "Integrating the Histories of Race and Technology," in *Technology and the African-American Experience: Needs and Opportunities for Study*, ed. Bruce Sinclair (Cambridge, MA: MIT Press, 2004), 5.

41. J. B. W., "Whittemore's Cotton Gin," *Southern Agriculturist* 8, no. 9 (September 1835): 475. Claims about the technological deficiency of Black workers were central to the marketing of tools "suited to the South." See Rockman, *Plantation Goods*, 211.

42. J. B. W., "Whittemore's Cotton Gin," *Southern Agriculturist* 8, no. 9 (September 1835): 475–476.

43. J. B. W., "Whittemore's Cotton Gin," 478.

44. J. B. W., "Whittemore's Cotton Gin," 478–479.

45. Lakwete, *Inventing the Cotton Gin*, 39.

46. For the molding of people to technologies, see Latour, "Where Are the Missing Masses?"

47. Awa Hanane Diagne et al., "The Black Androids: A Chrestomathy," *Historical Studies in the Natural Sciences* 54, no. 3 (2024): 420.

Chapter 4

1. See, e.g., Jennifer Karns Alexander, *The Mantra of Efficiency: From Waterwheel to Social Control* (Baltimore, MD: Johns Hopkins University Press, 2008), chap. 4; Peter B. Peterson, "Henry Gantt and the New Machine (1916–1919)," *Academy of Management Proceedings* 1986, no. 1 (1986): 128–132; Peter B. Peterson, "The Followers of Henry L. Gantt (1861–1919)," *Academy of Management Proceedings* 1993, no. 1 (1993): 174–178; and Daniel Nelson and Stuart Campbell, "Taylorism Versus Welfare Work in American Industry: H. L. Gantt and the Bancrofts," *Business History Review* 46, no. 1 (1972): 1–16. Samuel Haber mentions Gantt's family history and connection to the South but does not relate it to the development of scientific management. See Samuel Haber, *Efficiency and Uplift: Scientific Management in the Progressive Era, 1890–1920* (Chicago: University of Chicago Press, 1964), 42. Bill Cooke has claimed that this neglect of the place of slavery in the history of modern management practices amounts to "denial." See Bill Cooke, "The Denial of Slavery in Management Studies," *Journal of Management Studies* 40, no. 8 (2003): 1895–1918. Caitlin Rosenthal's excellent *Accounting for Slavery* focuses on this history in relation to Gantt's accounting practices rather than the Gantt chart itself. On histories of scientific management more generally, see, e.g., Haber, *Efficiency and Uplift*; Hugh G. J. Aitken, *Scientific Management in Action: Taylorism at Watertown Arsenal, 1908–1915* (Princeton, NJ: Princeton University Press, 1985); Martha Banta, *Taylored Lives: Narrative Productions in the Age of Taylor, Veblen, and Ford* (Chicago: University

of Chicago Press, 1993); Laurel D. Graham, "Domesticating Efficiency: Lillian Gilbreth's Scientific Management of Homemakers, 1924–1930," *Signs* 24, no. 3 (1999): 633–675; Daniel Nelson, *Frederick W. Taylor and the Rise of Scientific Management* (Madison: University of Wisconsin Press, 1980); Daniel Nelson, *A Mental Revolution: Scientific Management since Taylor* (Columbus: Ohio State University Press, 1992); Daniel Nelson, *Managers and Workers* (Madison: University of Wisconsin Press, 1995); and Robert Kanigel, *The One Best Way: Frederick Winslow Taylor and the Enigma of Efficiency* (New York: Viking, 1997).

2. Haber briefly discusses Gantt's view of idleness. See Haber, *Efficiency and Uplift*, 44. Alexander and Peterson focus instead on efficiency and do not mention the importance of idleness in Gantt's thought. See Alexander, *Mantra of Efficiency*, chap 4; Peterson, "Henry Gantt and the New Machine (1916–1919)"; and Peterson, "The Followers of Henry L. Gantt (1861–1919)."

3. On "paper technologies," see, e.g., Markus Krajewski, *Paper Machines: About Cards and Catalogs, 1548–1929*, trans. Peter Krapp (Cambridge, MA: MIT Press, 2011). For a more general discussion of "paper tools," see Ursula Klein, *Experiments, Models, Paper Tools: Cultures of Organic Chemistry in the Nineteenth Century* (Stanford, CA: Stanford University Press, 2003); David Kaiser et al., "Spreading the Tools of Theory: Feynman Diagrams in the USA, Japan, and the Soviet Union," *Social Studies of Science* 34, no. 6 (2004): 879–922; and Tresch, "Comte's Calendar: From Infinite Universe to Closed World," in *The Romantic Machine*, 253-286.

4. For the history of failure as a kind of man, see Scott A. Sandage, *Born Losers: A History of Failure in America* (Cambridge, MA: Harvard University Press, 2006).

5. I am deeply indebted to the work of Jackson Lears for these insights. See T. J. Jackson Lears, *Rebirth of a Nation: The Making of Modern America, 1877–1920* (New York: HarperCollins, 2009), introduction.

6. I draw here on Ruha Benjamin's insights regarding digital technologies and the "new Jim Code." See Benjamin, *Race After Technology*. According to Cedric Robinson, this differentiating process was central to the development of Western racial capitalism. See Cedric J. Robinson, *Black Marxism: The Making of the Black Radical Tradition* (Chapel Hill: University of North Carolina Press, 2000).

7. For the few studies of scientific management that focus on race, see Paul R. D. Lawrie, *Forging a Laboring Race: The African American Worker in the Progressive Imagination* (New York: New York University Press, 2016); Cooke, "The Denial of Slavery in Management Studies"; Banta, *Taylored Lives*; E. H. Brown, *The Corporate Eye: Photography and the Rationalization of American Commercial Culture, 1884–1929* (Baltimore, MD: Johns Hopkins University Press, 2008); and Rosenthal, *Accounting for Slavery*.

8. James Parvin Quigel Jr., "The Business of Selling Efficiency: Harrington Emerson and the Emerson Efficiency Engineers, 1900–1930" (PhD diss., Pennsylvania State University, 1992), 368–369. See also the proceedings of the Interstate Commerce

Commission, "Evidence Taken by the Interstate Commerce Commission in the Matter of Proposed Advances in Freight Rates by Carriers," August to December 1910, vol. 4, 61st Congress three-day session (Washington, DC, 1911), 2620.

9. Horace Bookwalter Drury, *Scientific Management: A History and Criticism* (New York: Columbia University Press, 1915), 27.

10. Lears, *Rebirth of a Nation*, 1–7.

11. Lears, *Rebirth of a Nation*, 2, 5, 11.

12. Lears, *Rebirth of a Nation*, 6.

13. Frederick Winslow Taylor, *Shop Management* (New York: Harper and Brothers, 1911), 190. On the broader Progressive Era notion of inborn qualities, see Sandage, *Born Losers*.

14. On the long-standing role of self-examination in Protestant conversion, see David W. Kling, "English Protestantism (1520–1700)," in *A History of Christian Conversion* (Oxford University Press, 2020), chap. 11, especially 287–288.

15. On Taylorist efficiency as a conversion experience, see Lears, *Rebirth of a Nation*, 265.

16. On the broader role of Protestantism in the Gilded Age and Progressive Era, see Lears, *Rebirth of a Nation*, 5–10.

17. Sandage, *Born Losers*, 9–10. See also Sarah E. Igo, *The Averaged American: Surveys, Citizens, and the Making of a Mass Public* (Cambridge, MA: Harvard University Press, 2007).

18. Theodore Roosevelt, "Fifth Annual Message," December 5, 1905.

19. "President Taft Talks for Wilberforcians," *Baltimore Afro-American* (March 12, 1910).

20. "Efficiency Prize." *Chicago Defender*, May 17, 1913.

21. See, e.g., Haber, *Efficiency and Uplift*, chap. 2.

22. Frederick Winslow Taylor, "A Piece-Rate System, Being a Step Toward Partial Solution of the Labor Problem," *Transactions of the American Society of Mechanical Engineers* 16 (1895), 856–903. Jennifer Alexander mistakenly attributes this system to Gantt, who merely implemented it. See Alexander, *Mantra of Efficiency*, 90–91.

23. On racial capitalism, see Robinson, *Black Marxism*. For a history and overview of the theory of racial capitalism and its place in Robinson's work, see Robin D. G. Kelley, foreword to Robinson, *Black Marxism*. For a collection of representative works, see Kelley, foreword to Robinson, *Black Marxism*, xxiv, n11.

24. Quoted in W. E. B. Du Bois, *The Philadelphia Negro: A Social Study* (Philadelphia: Published for the University, 1899; repr. University of Pennsylvania Press, 1996), 127.

25. Walter Licht, *Getting Work: Philadelphia, 1840–1950* (Philadelphia: University of Pennsylvania Press, 2000), 46, 109, 178.

26. W. E. B. Du Bois, *The Souls of Black Folk* (New York: Dover Publications, 1903), 2–3.

27. Rosenthal, *Accounting for Slavery*.

28. On the spread of time discipline to the plantation, see Mark M. Smith, *Mastered by the Clock: Time, Slavery, and Freedom in the American South* (Chapel Hill: University of North Carolina Press, 1997).

29. Rosenthal, *Accounting for Slavery*, 98. See also Cooke, "The Denial of Slavery in Management Studies"; and R. Keith Aufhauser, "Slavery and Scientific Management," *Journal of Economic History* 33, no. 4 (1973): 811–824. For Taylor's view on the task system, see Frederick Winslow Taylor, *The Principles of Scientific Management* (New York: Harper and Brothers, 1919), 39. On the details and history of the task system, see Faust, "In Search of Despotic Sway: Hammond as a Master of Slaves," in *James Henry Hammond and the Old South*; Chaplin, *An Anxious Pursuit*, 326–327; Joyner, *Down by the Riverside*, 43–46; Coclanis, "How the Low Country Was Taken to Task"; Morgan, "Task and Gang Systems"; and Morgan, *Slave Counterpoint*, 179–194.

30. Rosenthal, *Accounting for Slavery*, 201.

31. Gantt's early life is chronicled in Leon Pratt Alford, *Henry Laurence Gantt: Leader in Industry* (New York: American Society of Mechanical Engineers, 1934). Alford's biography is both a hagiography of Gantt and an apologia for slavery.

32. On the gendered history of idleness in Progressive Era America, see Daniel T. Rodgers, *The Work Ethic in Industrial America 1850–1920* (University of Chicago Press, 2014), chap. 7. On the role of urbanization in the construction of "idleness" and disability, see Sarah F Rose, *No Right to Be Idle: The Invention of Disability, 1840s–1930s* (Chapel Hill: University of North Carolina Press, 2017).

33. See, e.g., Maarten Derksen, "Turning Men into Machines? Scientific Management, Industrial Psychology, and the 'Human Factor,'" *Journal of the History of the Behavioral Sciences* 50, no. 2 (2014): 148–165.

34. See Aitken, *Scientific Management in Action*, 17. See also Daniel Nelson, "Scientific Management in Transition: Frederick W. Taylor at Johnstown, 1896," Pennsylvania Magazine of History and Biography 99, no. 4 (1975): 464; and Frederick Winslow Taylor, "On the Art of Cutting Metals," *Transactions of the American Society of Mechanical Engineers* 28, no. 2 (1906): 34–40.

35. Frederick W. Taylor, "Success: A Lecture to Young Men Entering Business," *Bulletin of the Taylor Society* 9 (1926): 74–75. Ironically, this lecture on "success," is filled with instances of breakdowns and machine failures.

36. Samuel Gompers, "The Miracles of 'Efficiency,'" *American Federationist* 18, no. 4 (1911): 279.

37. Kanigel, *The One Best Way*, 165. See also Horace B. Drury, *Scientific Management*, 2nd ed. (New York: Columbia University, 1918), 108–109.

38. Frederick Winslow Taylor, *Shop Management* (New York: Harper and Brothers, 1912), 66.

39. Taylor, *Shop Management*, 24.

40. For a similar arrangement in the British context, see M. Norton Wise, "The Gender of Automata in Victorian Britain," *Genesis Redux: Essays in the History and Philosophy of Artificial Life*, ed. Jessica Riskin (Chicago: University of Chicago Press, 2007), 168.

41. Kanigel, *The One Best Way*, 200–201.

42. Frederick W. Taylor, "Notes on Belting," *Transactions of the American Society of Mechanical Engineers*, 15 (1893): 212. See also Kanigel, *The One Best Way*.

43. Louis B. Brandeis, "Scientific Management and the Railroads," *Engineering Magazine*, 1911, 32.

44. Robert Cartwright, comment on Frederick W. Taylor's "Notes on Belting," *Transactions of the American Society of Mechanical Engineers*, 15 (1893), 252.

45. Taylor, "Notes on Belting," 213, 216.

46. United States Congress House Special Committee to Investigate the Taylor and Other Systems of Shop Management, *The Taylor and Other Systems of Shop Management. Hearings Before Special Committee of the House of Representatives to Investigate the Taylor and Other Systems of Shop Management Under Authority of H. Res. 90. [Oct. 4, 1911–Feb. 12, 1912]* Vol. 3 (Washington: Government Printing Office, 1912), 1414.

47. A similar system of labor, punishment, and identification with breaking machines existed in the spinning factories of Dasheng, China. See Yuang, "Malfunctioning Machinery," chap. 4.

48. Alford, *Henry Laurence Gantt*, 51. For another example, see Banta, *Taylored Lives*, 127.

49. Peterson, "The Followers of Henry L. Gantt (1861–1919)," 174.

50. On the question of industrial espionage, see Charles D. Wrege and Ronald G. Greenwood, "Frederick W. Taylor and Industrial Espionage: 1895–1897," *Business and Economic History* 15 (1986): 183–193.

51. See Daniel Nelson, "Taylorism and the Workers at Bethlehem Steel, 1898–1901," *Technology and Culture* 101, no. 4 (1977): 487–505.

52. Henry Laurence Gantt, "A Bonus System of Rewarding Labor, Being a System of Task, with Instruction Cards and a Bonus," *Transactions of the American Society*

of Mechanical Engineers XXIII (1901), 351. As Gantt acknowledged, this was based on Taylor's system of elementary rate fixing. See Taylor, "A Piece-Rate System."

53. Gantt, "A Bonus System of Rewarding Labor," 343, 358 (emphasis added).

54. Letter from Henry Laurence Gantt to Frederick Winslow Taylor, April 25, 1913. Frederick Winslow Taylor collection (hereafter FWT collection), ms.001.36424.

55. Gantt had already developed a daily balance chart for his task-and-bonus system at the American Locomotive Company. See Alford, *Henry Laurence Gantt*, 208.

56. John Golden, "'Scientific Management' in the Textile Industry," *American Federationist* 18, no. 8 (1911): 603–604.

57. Sue Ainslie Clark and Edith Wyatt, *Making Both Ends Meet: The Income and Outlay of New York Working Girls* (New York: Macmillan Company, 1911), 251.

58. Gantt to Taylor, April 7, 1904. FWT collection, ms.001.35339.

59. Henry Laurence Gantt, "Work, Wages, and Profits," *Engineering Magazine*, 1910, 145.

60. Brandeis, *Scientific Management and the Railroads*, 37 (emphasis added).

61. Gantt, "Work, Wages, and Profits," 181.

62. Golden, "'Scientific Management' in the Textile Industry," 603. In 1911, Gantt and Taylor were under investigation by a House Special Committee in light of the Watertown Arsenal Strike. They mounted a public relations campaign in response, and Taylor attempted to avoid testifying. See Letter from Frederick Winslow Taylor to Henry Laurence Gantt, October 5, 1911, FWT collection ms.001.36336.

63. Clark and Wyatt, *Making Both Ends Meet*, 246.

64. Golden, "'Scientific Management' in the Textile Industry," 604.

65. Clark and Wyatt, *Making Both Ends Meet*, 246. The text refers to a "New England Cloth Finishing House," but the context and description identify it as Sayles Bleacheries.

66. Henry Laurence Gantt, "Scheduling Locomotive Repair Work on the Canadian Pacific Railway," *Industrial Engineering and the Engineering Digest* 8, no. 5 (1910): 380.

67. Editorial Correspondence, "Canadian Pacific Shop Management," *American Machinist* 35 (December 21, 1911): 1164.

68. The Congressional investigation of scientific management stemmed largely from the 1911 strike at the Watertwon arsenal. See Daniel T. Rodgers, *The Work Ethic in Industrial America, 1850-1920*, 2nd ed. (Chicago: The University of Chicago Press, 2014), 166-173. On the splintering of the movement, see Nelson, *Managers and Workers*, 67,

73–74. The findings of the House Special Committee were published under "Hearings before the Special Committee of the House of Representatives to Investigate the Taylor and Other Systems of Shop Management" (Washington, DC, 1912).

69. Henry Laurence Gantt, *Organizing for Work* (New York, Harcourt, Brace and Howe, 1919), 96.

70. See Peterson, "Henry Gantt and the New Machine (1916–1919)." On the composition of the New Machine and the influence of Thorstein Veblen, see Haber, *Efficiency and Uplift*, 44–49; and Charles Ferguson, *Philadelphia Public Ledger*, December 10, 1916. The New Machine's elevation of engineers would mirror Thorstein Veblen's *Engineers and the Price System* (New York: B. W. Huebsch, 1921). I'm grateful to Kenneth Lipartito for drawing this connection.

71. Charles Ferguson, *The Revolution Absolute* (New York: Dodd, Mead, 1918), 278, see 301–311 for the letter to the president.

72. Ferguson, *The Revolution Absolute*, vii-I, 272–279, 301–311.

73. Peterson, "The Followers of Henry L. Gantt (1861–1919)," 176–177.

74. Henry Laurence Gantt, "Efficiency and Idleness," *Industrial Management* 56, no. 5 (1918): 377; and Gantt, *Organizing for Work*, 25.

75. Gantt, *Organizing for Work*, 26.

76. Quoted in Gantt, *Organizing for Work*, 107.

77. Gantt, *Organizing for Work*, 25.

78. See Joe William Trotter Jr., ed., *The Great Migration in Historical Perspective: New Dimensions of Race, Class, and Gender* (Bloomington: Indiana University Press, 1991); Nicholas Lemann, *The Promised Land: The Great Black Migration and How It Changed America* (New York: Vintage, 1992); and James Gregory, *The Southern Diaspora: How the Great Migrations of Black and White Southerners Transformed America* (Chapel Hill: University of North Carolina Press, 2005). See also Isabel Wilkerson, *The Warmth of Other Suns: The Epic Story of America's Great Migration* (New York: Vintage, 2011).

79. Philip H. Brown, "The Negro Migrant," *Bulletin of the U.S. Bureau of Labor Statistics*, no. 355 (March 1924): 47–52.

80. Lawrie, *Forging a Laboring Race*, chap. 2.

81. Daniel E. Bender, *American Abyss: Savagery and Civilization in the Age of Industry* (Ithaca, NY: Cornell University Press, 2013), chap. 1.

82. See Lawrie, *Forging a Laboring Race*, introduction; Bender, *American Abyss*, 21–22; and Banta, *Taylored Lives*, 4.

83. On the related concept of otium in classical Roman and Renaissance culture, see Brian Vickers, "Leisure and Idleness in the Renaissance: The Ambivalence of Otium,"

Renaissance Studies 4, no. 1 (1990): 1–37. On the Christian concept of acedia, see Siegfried Wenzel, *The Sin of Sloth: Acedia in Medieval Thought and Literature* (Chapel Hill: University of North Carolina Press, 1960). Friedrich Nietzsche, in contrast, would praise idleness as a virtue in *Menschliches, Allzumenschliches: Ein Buch für freie Geister* (Chemnitz: Verlag von Ernst Schmeitzner, 1878), 284. On medieval idleness, see Jacques Le Goff, "Labor Time in the 'Crisis' of the Fourteenth Century: From Medieval Time to Modern Time," in *Time, Work, and Culture in the Middle Ages*, trans. Arthur Goldhammer (Chicago: Chicago University Press, 1980), 51–52. For a fascinating study of idleness in the late Ottoman Empire, see Avner Wishnitzer, *Reading Clocks, Alla Turca: Time and Society in the Late Ottoman Empire* (Chicago: University of Chicago Press, 2015), chap.3. On Walter Benjamin's association of idleness with the triumph of modernity, see Walter Benjamin, *Das Passagen-Werk. Gesammelte Schrifte*, vol. 5:2 [*Arcades Project*] (Frankfurt: Suhrkamp, 1982), 962. On the longer history of idleness and its supersession by "fatigue" in the nineteenth-century European sciences of work, see Rabinbach, *The Human Motor*.

84. Thompson, "Time, Work-Discipline, and Industrial Capitalism," 81. On idleness as the legacy of the preindustrial and the trait of the "uncivilized," see Rabinbach, *The Human Motor*, 28–29.

85. Rodgers, *Work Ethic in Industrial America*, chap. 7. On the gendered associations of busyness, see Michael O'Malley, "That Busyness That Is Not Business: Nervousness and Character at the Turn of the Last Century," *Social Research* 72, no. 2 (2005): 371–406.

86. Sandage, *Born Losers*, 203. See also Rodgers, *Work Ethic in Industrial America*, 167.

87. Henry Laurence Gantt, "The Religion of Democracy," *Industrial Management* 57, no. 5 (1919): 405.

88. Gantt, *Organizing for Work*, 86–88.

89. Gantt's bonus scheme was an elaboration of "premium" plans popularized by Frederick W. Halsey. See Frederick A. Halsey, "The Premium Plan of Paying for Labor," *Transactions of the American Society of Mechanical Engineers* 22 (1891): 758. See also Nelson, "Scientific Management in Transition," 463.

90. For a more general discussion of this transformation within the racial politics of the Gilded Age and Progressive Era, see Khalil Gibran Muhammad, *The Condemnation of Blackness: Race, Crime, and the Making of Modern Urban America* (Cambridge, MA: Harvard University Press, 2010); and James R. Barrett, "Americanization from the Bottom Up: Immigration and the Remaking of the Working Class in the United States, 1880–1930," *Journal of American History* 79, no. 3 (1992): 996–1020.

91. Licht, *Getting Work*, 45.

92. Jacqueline Jones, "'Lifework' and Its Limits: The Problem of Labor in *The Philadelphia Negro*," in *W. E. B. Du Bois, Race, and the City: The Philadelphia Negro and Its*

Legacy, ed. Michael B Katz, and Thomas J Sugrue (Philadelphia: University of Pennsylvania Press, 1998), 110.

93. On the afterlives of slavery, see Saidiya Hartman, *Lose your Mother: A Journey Along the Atlantic Slave Trade Route* (New York: Farrar, Strauss, and Giroux, 2007), 6.

94. Lawrie, *Forging a Laboring Race*, 55.

95. Lawrie, *Forging a Laboring Race*, 56.

96. Quoted in George E. Haynes, *The Negro at Work During the World War and During Reconstruction: Statistics, Problems, and Policies Relating to the Greater Inclusion of Negro Wage Earners in American Industry and Agriculture* (Washington, DC: Government Printing Office, 1921), 62 and 63.

97. Quoted in Lawrie, *Forging a Laboring Race*, 56, 187, n75.

98. Quoted in Lawrie, *Forging a Laboring Race*, 62, n97

99. Quoted in Lawrie, *Forging a Laboring Race*, 63, n98.

100. Henry L. Gantt, "Democratic Shop Methods," Speech, De Laval Steam Turbine Co., January 29, 1919; Henry L. Gantt Papers, New York University.

101. Gantt, "Democratic Shop Methods."

102. "Program Ninth Annual Session Southern Commercial Congress, New York October 15, 16 and 17," *Southern Good Roads*, October 1917. See also Lawrie, *Forging a Laboring Race*, 51.

103. Ethelene Whitmire, *Regina Anderson Andrews, Harlem Renaissance Librarian* (Urbana: University of Illinois Press, 2014), 51–53.

104. See Janet G. Hudson, *Entangled by White Supremacy: Reform in World War I–Era South Carolina* (Lexington: University Press of Kentucky, 2009), 154–155.

105. *Nashville Globe*, February 23, 1917.

106. *Nashville Globe*, February 23, 1917.

107. Gantt's vision is particularly vicious, since, as Joel Dinerstein has argued, African American culture had arguably done more to integrate the rhythms of the machine into early twentieth-century American life than any other. See Dinerstein, *Swinging the Machine*.

108. For a similar perspective on digital technologies and dystopias, see Benjamin, *Race After Technology*, chap. 3.

109. Jodi Melamed, "Racial Capitalism," *Critical Ethnic Studies* 1, no. 1 (2015): 77. See also Robinson, *Black Marxism*; Destin Jenkins and Justin Leroy, eds., *Histories of Racial Capitalism* (New York: Columbia University Press, 2021); and Keeanga-Yamahtta Taylor, *From #BlackLivesMatter to Black Liberation* (Chicago: Haymarket, 2016).

110. On the broader entanglements of race and technology, see, e.g., chapter 1, footnote 37.

Chapter 5

1. D. E. Noble quoted in "Noble Sees Reliability Failure Limiting War Role of ICBMs," *Electronics News* (November 21, 1960): 43. On technological failures during the Cold War, see, e.g., Eric Schlosser, *Command and Control: Nuclear Weapons, the Damascus Accident, and the Illusion of Safety* (New York: Penguin, 2013); and Donald MacKenzie, *Inventing Accuracy: A Historical Sociology of Nuclear Missile Guidance* (Cambridge, MA: MIT Press, 1993). On later risks of catastrophic computer failure, see Slayton, *Arguments That Count.*

2. A notable exception here is Layne Karafantis and Daniel Volmar, "Idiot-Proofing the Air Force: Human Factors Engineering and the Crisis of Systems of Maintenance," paper presented at *The Maintainers II Labor: Technology, and Social Orders*, April 6–9, 2017, Hoboken, NJ.

3. See, e.g., Michael Riordan, and Lillian Hoddeson, *Crystal Fire: The Birth of the Information Age* (New York: W. W. Norton & Company, 1997); Christophe Lécuyer, *Making Silicon Valley: Innovation and the Growth of High Tech, 1930–1970* (Cambridge, MA: MIT Press, 2007); and Margaret O'Mara, *The Code: Silicon Valley and the Remaking of America* (New York: Penguin, 2020). For a critical contemporary perspective, see Mary Beth Meehan and Fred Turner, *Seeing Silicon Valley: Life Inside a Fraying America* (Chicago: University of Chicago Press, 2021).

4. On Shockley's place in this history, see the excellent article by David C. Brock, "From Automation to Silicon Valley: The Automation Movement of the 1950s, Arnold Beckman, and William Shockley," *History and Technology* 28, no. 4 (2012): 375–401.

5. Robert Lusser, "Unreliability of Electronics: Cause and Cure" (Huntsville, AL: Research and Development Division, Ordnance Missile Laboratories, Redstone Arsenal, November 1957).

6. Rsystem = R1 × R2 × R3 × . . . Rn. The expression assumes that (a) the individual part reliabilities are independent, (b) they are arranged in series, and (c) a failure of any part will fail the equipment (i.e., each is critical to operation). For an expression of this view in relation to the "automation crisis," see George Steele and Paul Kircher, *The Crisis We Face: Automation and the Cold War* (New York: McGraw-Hill, 1960), 127–129.

7. Lusser, "Unreliability of Electronics," 5.

8. Lusser, "Unreliability of Electronics," 7.

9. Chester I. Soucy, "Reliability Control of Electronic Equipment in Aircraft and Weapons Systems: Its General and Management Aspects," *Canadian Aeronautical Journal* 3, no. 7 (September 1957): 223.

10. For archival materials related to Soucy, see University of Toronto Archives, ref.: A1998-0003 Reel 008.

11. Robert Lusser, "The Notorious Unreliability of Complex Equipment: A Condensed Presentation for Management Use" (Huntsville, AL: Research and Development Division, Ordnance Missile Laboratories, Redstone Arsenal, 1956).

12. Quoted in Lusser "Unreliability of Electronics," 2.

13. Brigadier General J. M. Colby, "Minutes of the Second AOMC Agencies Conference Regarding Guided Missile Reliability," held at US Army Missile Command, Huntsville, AL (May 12, 1959), p. 1; Library and Archives Canada (hereafter LAC) RG 24, accession 1983-84/049, vol. 1664, file 1950-123-7, part 2.

14. See, e.g., Royal Canadian Air Force, *Reliability Bulletin*, nos. 1 and 2, i.; LAC RG 24, accession 83-84/049, vol. 1664, file 1950-123-7, part 2.

15. Chester I Soucy, "A Perspective View of Electronic Component Part Reliability," in *Proceedings of the Canadian Military Electronic Components Symposium*, April 11–12, 1961, Ottawa, Canada, 6 (emphasis added).

16. See Jamie Cohen-Cole, *The Open Mind: Cold War Politics and the Sciences of Human Nature* (Chicago: University of Chicago Press, 2016), 36–39, 115. See also Alan Brinkley, "The Illusion of Unity in Cold War Culture," in *Rethinking Cold War Culture*, ed. Peter J. Kuznick and James Gilbert (Washington, DC: Smithsonian Institution Press, 2001); Daniel Bell, *The End of Ideology: On the Exhaustion of Political Ideas in the Fifties* (New York: Free Press, 1960), 21–38; Talcott Parsons, "Evolutionary Universals in Society," *American Sociological Review* 29, no. 3 (June 1964): 339–357; Margaret Mead, *Coming of Age in Samoa: A Psychological Study of Primitive Youth for Western Civilization* (New York: HarperCollins, 1928); and George F. Kennan, "America's Administrative Response to Its World Problems," in *The American Style: Essays in Value and Performance*, ed. Elting E. Morison (New York: Harper and Brothers, 1957). For the prewar anxieties about modern complexity and American society, see Igo, *The Averaged American*.

17. See, e.g., Aaron Mendon-Plasek, "Mechanized Significance and Machine Learning: Why It Became Thinkable and Preferable to Teach Machines to Judge the World," in *The Cultural Life of Machine Learning*, ed. Jonathan Roberge and Michael Castelle (Cham: Springer International Publishing, 2021), 57.

18. Cohen-Cole, *The Open Mind*, chap. 2.

19. Here, I combine concerns over reason and rationality that historians have treated as separate and competing. On the triumph of Cold War rationality over reason, see Paul Erickson et al., *How Reason Almost Lost Its Mind: The Strange Career of Cold War Rationality* (Chicago: University of Chicago Press, 2013); Fred Kaplan, *The Wizards of Armageddon* (Stanford: Stanford University Press, 1991); and Sharon Ghamari-Tabrizi, *The Worlds of Herman Kahn: The Intuitive Science of Thermonuclear*

War (Cambridge, MA: Harvard University Press, 2005). On the continued flourishing of reason outside centers such as RAND, see Jamie Cohen-Cole, "Reason Beyond RAND: Did Enlightenment Values Persist Among Cold War Intellectuals?," *History of the Human Sciences* 31 (2017): 122–126.

20. See Edward Jones-Imhotep, "Paris-Montreal-Babylon: The Modernist Genealogies of Gerald Bull," in *Made Modern: Science and Technology in Canadian History*, ed. Edward Jones-Imhotep and Tina Adcock (Vancouver: UBC Press, 2018). For the development of "assured destruction" and second-strike doctrine, see James Lee and David Bellamy, "Dr R. J. Sutherland: A Retrospective by James Lee and David Bellamy," Directorate of History and Heritage of the Department of National Defence, collection 87/253 III, box 6, file 0, 1; and "A Military View of Nuclear Weapons," February 28, 1961, LAC RG 24, accession 1983-84/167, box 7373, DRBS 170-80/J56, vol. 3.

21. Joy Rohde, "Pax Technologica: Computers, International Affairs, and Human Reason in the Cold War," *Isis* 108, no. 4 (2017): 794.

22. See the excellent discussion of this point in Hunter Heyck, *Age of System: Understanding the Development of Modern Social Science* (Baltimore, MD: Johns Hopkins University Press, 2015), chap. 4. This project built on a longer tradition of twentieth-century social thinkers who had already replaced the nineteenth-century notions of the self—the freestanding, autonomous product of will and ambition—with a less self-standing entity, encircling the self "with wider and wider rings of relations, structures, contexts, and institutions." See Daniel T. Rodgers, *Age of Fracture* (Cambridge, MA: Harvard University Press, 2011), 4.

23. The historical literature on cybernetics is vast. See. e.g., Edwards, *The Closed World*, 75–111; Peter Galison, "Ontology of the Enemy: Norbert Wiener and the Cybernetic Vision," *Critical Inquiry* 21 (1994): 228–266; Hayles, *How We Became Posthuman*; Steve J. Heims, *Constructing a Social Science for Postwar America: The Cybernetics Group, 1946–1953* (Cambridge, MA: MIT Press, 1993); Ronald L. Kline, *The Cybernetics Moment; or, Why We Call Our Age the Information Age* (Baltimore, MD: Johns Hopkins University Press, 2015); Andrew Pickering, *The Cybernetic Brain: Sketches of Another Future* (Chicago: University of Chicago Press, 2010); Eden Medina, *Cybernetic Revolutionaries: Technology and Politics in Allende's Chile* (Cambridge, MA: MIT Press, 2011); and Slava Gerovitch, *From Newspeak to Cyberspeak: A History of Soviet Cybernetics* (Cambridge, MA: MIT Press, 2004).

24. On the ongoing breakdowns in navy equipment, see R. R. Carhart, "A Survey of the Current Status of the Electronic Reliability Problem," US Air Force Project RAND Research Memorandum RM-1131 (Santa Monica, 1953). Carhart suggested that up to two-thirds of Western military equipment might be malfunctioning. See Carhart, "A Survey of the Current Status of the Electronic Reliability Problem," 66.

25. Glenn L. Bryan, "Training of Electronic Maintenance Technicians," *Naval Research Reviews*, 1960–11: 9.

26. The Group's director, Glenn Bryan, would go on to head the Psychological Sciences Program at the Office of Naval Research for twenty years. His group deeply influenced the military's efforts to cope with maintenance in complex systems. See Stanley N. Roscoe, *The Adolescence of Engineering Psychology* (Santa Monica, CA: Human Factors and Ergonomics Society, 1997), 6. For a later, similarly widescale initiative to create a worldwide incidence response network for the Internet, see Rebecca Slayton and Brian Clarke, "Trusting Infrastructure: The Emergence of Computer Security Incident Response, 1989–2005," *Technology and Culture* 61, no. 1 (2020): 173–206.

27. Edgerton, *The Shock of the Old*, 79. See also Jackson, "Rethinking Repair"; Graham and Thrift, "Out of Order," 1–2; and Stefan Krebs and Heike Weber, eds., *The Persistence of Technology: Histories of Repair, Reuse and Disposal* (Bielefeld, Germany: Transcript Verlag, 2021).

28. For example see Larkin, *Signal and Noise*; Jojada Verrips and Brigit Meyer, "Kwaku's Car: The Struggles and Stories of a Ghanaian Long-Distance Taxi-Driver," in *Car Cultures*, ed. Daniel Miller (Oxford: Berg, 2001); John Powell, *The Survival of the Fitter: Lives of Some African Engineers* (London: Intermediate Technology, 1995); Joshua Grace, *African Motors: Technology, Gender, and the History of Development* (Durham, NC: Duke University Press, 2021); Tasha Rijke-Epstein, "On Humble Technologies: Containers, Care, and Water Infrastructure in Northwest Madagascar, 1750s–1960s," *History and Technology* 37, no. 3 (2021): 293–328; Burrell, *Invisible Users*; and the special thematic cluster, Raquel Velho and Sebastián Ureta, "Frail Modernities: Latin American Infrastructures Between Repair and Ruination," *Tapuya* 2, no. 1 (2019): 428–459.

29. Russell and Vinsel, "After Innovation, Turn to Maintenance."

30. It is worth noting that "San" is a derogatory Khoekhoegowab term imposed on distinct groups of Southern Africa who do not identify themselves by a common name.

31. Russell and Vinsel, "After Innovation, Turn to Maintenance," 7.

32. One of the very few works to historicize maintenance in this moment is Karafantis and Volmar, "Idiot-Proofing the Air Force." For works historicizing breakage, repair, and maintenance, see Schaffer, "Easily Cracked"; Pierre Claude Reynard, "Unreliable Mills: Maintenance Practices in Early Modern Papermaking," *Technology and Culture* 40, no. 2 (1999): 237–262; and Kevin L. Borg, *Auto Mechanics: Technology and Expertise in Twentieth-Century America* (Baltimore, MD: Johns Hopkins University Press, 2007). For an attempt to differentiate maintenance practices, see Jérôme Denis and David Pontille, "Beyond Breakdown: Exploring Regimes of Maintenance," *Continent* 6, no. 1 (2017): 13–17.

33. Researchers observed that the likelihood that a random electronic part would fail was high during its initial "burn-in" period, then dropped dramatically before

rising again toward the end of their lifespans. See Keith Henney, ed., *Reliability Factors for Ground Electronic Equipment* (New York: McGraw-Hill, 1956), 4–5.

34. Memo "Electronics Reliability Programme Reliability Bulletins," September 29, 1958, nos. 1 and 2, i.; LAC RG 24, accession 1983-84/049, vol. 1664, file 1950-123-7, part 2.

35. Henney, *Reliability Factors for Ground Electronic Equipment*, 1–2.

36. On maintenance as a "manpower" problem, see Chester I. Soucy, "A Broad Survey of the Military Electronic Equipment Reliability Problem," *Proceedings, 1956 Electronic Components Symposium* (New York: Engineering Publishers, 1956), 8. On the wider "manpower" shortages of the period, see Slayton, *Arguments that Count*; and David Kaiser, "Cold War Requisitions, Scientific Manpower, and the Production of American Physicists After World War II," *Historical Studies in the Physical and Biological Sciences* 33, no. 1 (2002): 131–159.

37. Glenn L. Bryan, "Electronics Trouble Shooting: A Behavioral Analysis," Technical Report No. 13, Department of Psychology, University of Southern California, LA, March 1956, 166–167.

38. For the contemporary interest in teaching machines for maintenance, see Eugene Galanter, ed., *Teaching Machines for Training of Military Personnel in Maintenance of Electronic Equipment* (New York: Wiley, 1959).

39. Bryan, "Electronics Trouble Shooting," 121–123; and Bryan, "Training of Electronic Maintenance Technicians," 11–12.

40. Audrey Watters, *Teaching Machines: The History of Personalized Learning* (Cambridge, MA: MIT Press, 2021). The Soviet Union, the United Kingdom, and West Germany all experimented with the devices. See Watters, *Teaching Machines*, 157. For the Soviet case and the PLATO initiative at the University of Illinois, see Ekaterina Igorevna Babintseva, "Cyberdreams of the Information Age: Learning with Machines in the Cold War United States and the Soviet Union" (PhD diss., University of Pennsylvania, 2020).

41. Stephen Petrina, "Sidney Pressey and the Automation of Education, 1924–1934," *Technology and Culture* 45, no. 2 (2004): 305–330. See also Oscar H. Gandy, "Instructional Technology: The Reselling of the Pentagon" (PhD diss., Stanford University, 1976), 111–112.

42. See Watters, *Teaching Machines*, chap. 4; and Petrina, "Sidney Pressey and the Automation of Education, 1924–1934."

43. Robert Czeh, for instance, longtime friend of historian of technology, Thomas P. Hughes, drew explicit comparisons between electronics troubleshooting approaches and Dewey's model of thinking expressed in *How We Think* (Boston: D.C. Heath & Co., 1910). See Robert Czeh, "Studies of Complex Behavior and Their Relation

to Trouble Shooting in Electronic Equipment," US Army Contract DA-49-105-qm-1 (Alexandria, VA: Human Resources Research Organization, 1957), 3.

44. On Dewey's communities, see Saraiva, *The Orchard in Ruins*; and Cornel West, *American Evasion of Philosophy: A Genealogy of Pragmatism* (Madison: University of Wisconsin Press, 1989), 106.

45. On the racialized character of universals, see, e.g., Mills, *The Racial Contract*. For an excellent analysis of the conflation of racial aptitude and technical ability, see Slaton, *Race, Rigor, and Selectivity in US Engineering*.

46. Light, "When Computers Were Women," 469. See also Janet Abbate, *Recoding Gender: Women's Changing Participation in Computing* (Cambridge, MA: MIT Press, 2012), 29.

47. Mar Hicks, *Programmed Inequality: How Britain Discarded Women Technologists and Lost Its Edge in Computing* (Cambridge, MA: MIT Press, 2017).

48. On the feminized (and racialized) work of electronic production, see Light, "When Computers Were Women"; and Lisa Nakamura, "Indigenous Circuits: Navajo Women and the Racialization of Early Electronic Manufacture," *American Quarterly* 66, no. 4 (2014): 919–941. On the role of technologies in shaping gender and notions of "technical" affinity, see, e.g., Wajcman, *Technofeminism*; Bray, "Gender and Technology"; and Lerman, "Categories of Difference, Categories of Power."

49. Paul Baran noted that in complex systems such as the "survivable" relay networks he was developing, "intelligence," particularly logic, was required "to determine where failure has occurred." See Paul Baran, *Reliable Digital Communications Systems Using Unreliable Network Repeater Nodes* (Santa Monica, CA: RAND Corporation, 1960).

50. Soucy, "A Perspective View of Electronic Component Part Reliability," 5.

51. Steele and Kircher, *The Crisis We Face*, 194.

52. Rebecca Lemov, "Hypothetical Machines: The Science Fiction Dreams of Cold War Social Science," *Isis* 101, no. 2 (2010): 406–407.

53. Ken Alder, "America's Two Gadgets: Of Bombs and Polygraphs," *Isis* 98, no. 1 (2007): 130–131.

54. Chester I. Soucy, "Aerospace-System Applications of Reliability Technology Derived from the Electronics Field," in *Proceedings of the 4th Congress of the International Council of the Aeronautical Sciences*, ed. Robert R. Dexter (Washington, DC: Spartan Books, 1964), 776.

55. Joel Cooper, "Human-Initiated Failures and Malfunction Reporting," *IRE Transactions on Human Factors in Electronics* HFE-2, no. 2 (1961): 107.

56. Soucy, "A Perspective View of Electronic Component Part Reliability," 6.

57. On DRB's role in Cold War nuclear strategy, see Directorate of History and Heritage, Col 87/253 III (Box 6) File 0, "Dr RJ Sutherland: A Retrospective by James Lee and David Bellamy," 1.

58. For DRB links to the NSA, see Jones-Imhotep, *The Unreliable Nation*, chap. 6. For general histories of DRB, see D. J. Goodspeed, *A History of the Defence Research Board of Canada* (Ottawa: Queen's Printer, 1958); and Jonathan Turner, "The Defence Research Board of Canada, 1947 to 1977" (PhD diss., University of Toronto, 2012). On the "hostile environments" of the Cold War, see Matthew Farish, "The Lab and the Land: Overcoming the Arctic in Cold War Alaska," *Isis* 104, no. 1 (2013): 1–29, and *The Contours of America's Cold War* (Minneapolis: University of Minnesota Press, 2010).

59. N. F. Moody, "The Present State of the Transistor and Its Associated Circuit Art," *Nuclear Instruments* 2, no. 2 (1958): 182–192. The historical literature on transistors is almost completely silent on the question of symbols. Linda Petiot acknowledges the existence of different conventions in her excellent study of Moody's computer research but remains silent on its significance. See Linda Petiot, "Dirty Gertie: The DRTE Computer," *IEEE Annals of the History of Computing* 16, no. 2 (1994): 43–52. For a contemporary overview during the Reliability Crisis, see Marcus Scroggie, "Transistor Graphic Symbols: A Critical Analysis of Existing Ideas and Conventions," *Wireless World* 63 (1957): 194–198.

60. For a fuller discussion of these proposals and the issues they were designed to solve, see Edward Jones-Imhotep, "Icons and Electronics," *Historical Studies in the Natural Sciences* 38, no. 3 (2008): 405–50.

61. For an early and influential articulation of this view, see Bruno Latour, "Drawing Things Together," in *Representation in Scientific Practice*, ed. Michael Lynch and Steve Woolgar (Cambridge, MA: MIT Press, 1990).

62. Alder, "Making Things the Same"; and David Kaiser, *Drawing Theories Apart: The Dispersion of Feynman Diagrams in Postwar Physics* (Chicago: University of Chicago Press, 2005).

63. See Wolfgang Lefevre, *Picturing Machines 1400–1700* (Cambridge, MA: MIT Press, 2004); Michael E. Gorman and W. Bernard Carlson, "Interpreting Invention as a Cognitive Process: The Case of Alexander Graham Bell, Thomas Edison, and the Telephone," *Science, Technology, and Human Values* 15, no. 2 (1990): 131–164; Eugene S. Ferguson, *Engineering and the Mind's Eye* (Cambridge, MA: MIT Press, 1994); and Roland Wittje, "The Electrical Imagination: Sound Analogies, Equivalent Circuits, and the Rise of Electroacoustics, 1863–1939," *Osiris* 28, no. 1 (2013): 40–63.

64. Here, I borrow the term "paper tool" from Ursula Klein and David Kaiser. See Klein, *Experiments, Models, Paper Tools*; and Kaiser, *Drawing Theories Apart*, chap. 1.

65. L. H. Bainbridge-Bell, "The Need for Clearer Circuit Diagrams," *Engineering* 170 (1950): 260.

66. Bainbridge-Bell, "The Need for Clearer Circuit Diagrams."

67. On manners and morals, Sarah Buss has explained that although most philosophers treat the two separately, most people draw no sharp distinction between them. See Sarah Buss, "Appearing Respectful: The Moral Significance of Manners," *Ethics* 109, no. 4 (1999): 796. For the stress on manners in the debate over symbols, see Henry Morgan, "Letter to the Editor," *Wireless World* 60 (1954): 178.

68. On the transistor as a replacement for vacuum tubes, see Stuart Macdonald and Ernest Braun, "The Transistor and Attitude Towards Change," *American Journal of Physics* 45, no. 11 (1977): 1062–1063; Stuart W. Leslie, "Blue Collar Science: Bringing the Transistor to Life in the Lehigh Valley," *Historical Studies in the Physical and Biological Sciences* 32, no. 1 (2001): 89–90; Riordan and Hoddeson, *Crystal Fire*, 5, n2; Gerald L. Pearson and Walter H. Brattain, "History of Semiconductor Research," *Proceedings of the IRE* 43, no. 12 (1955): 1794–1806; and Walter H. Brattain, "Genesis of the Transistor," *The Physics Teacher* 6, no. 3 (1968): 109–114.

69. Baran, *Reliable Digital Communications Systems*; and Parsons, "Evolutionary Universals in Society."

70. See Heyck, *Age of System*, 1. On the relation between social scientific theory and Cold War technology, see Heyck, *Age of System*, 42.

71. P. M. Thompson, "The Graphic Nomenclature of Transistors," EL Memorandum no. 5059–50A, Defence Research Board Telecommunications Establishment, Electronics Laboratory, Nov. 1954, 1, LAC RG 24, accession 89-90/205, box 91, file 170-180/T8, part 1.

72. Moody, "The Present State of the Transistor," 182.

73. Moody, "The Present State of the Transistor," 182.

74. Moody, "The Present State of the Transistor," 191. The historian Eugene Ferguson expressed a similar position about engineering judgment as the source of disaster. See Ferguson, *Engineering and the Mind's Eye*, 193.

75. N. F. Moody and C. D. Florida, "Some New Transistor Bistable Elements for Heavy Duty Operation," *IRE Transactions on Circuit Theory* 4, no. 3 (1957): 241.

76. David Mindell discusses this feature of the image in David A. Mindell, *Between Human and Machine: Feedback, Control, and Computing before Cybernetics* (Baltimore, MD: Johns Hopkins University Press, 2004), 69.

77. See, e.g., Neda Atanasoski and Kalindi Vora, *Surrogate Humanity* (Durham, NC: Duke University Press, 2019); Lilly Irani, "The Cultural Work of Microwork," *New Media & Society* 17, no. 5 (2013): 720–739; and Meehan and Turner, *Seeing Silicon Valley*.

78. I adapt this formulation from Simon Schaffer and Stephen Shapin's claim about knowledge and social order. See Shapin and Schaffer, *Leviathan and the Air-Pump*,

332. For an excellent discussion of this relation in another Cold War context, see Mendon-Plasek, "Mechanized Significance and Machine Learning,", 31–78.

Epilogue

1. See Tresch, "Technological World-Pictures," 84, 87.

2. On Heidegger's relationship to National Socialism, see Wolin, *Heidegger in Ruins*.

3. In this way, it functioned very closely to the bourgeois discourses of self-mastery discussed in Ann Stoler, *Race and the Education of Desire: Foucault's History of Sexuality and the Colonial Order of Things* (Durham, NC: Duke University Press, 1995), 2.

4. On the value of affective differences, see Sara Ahmed, "Affective Economies," *Social Text* 22, no. 2 (Summer 2004): 119–120.

5. Lewis Mumford, *The Golden Day* (New York: Horace Liverlight, 1926). Ralph Ellison, who acknowledged admiring Mumford, mounted a devastating critique of *The Golden Day* in *Invisible Man* (New York: Random House, 1952). Alan Nadell has suggested that Mumford's interpretation of the Civil War was a displaced reaction to World War I. For an analysis of Mumford's text and Ellison's response, see Alan Nadel, *Invisible Criticism: Ralph Ellison and the American Canon* (Iowa City: University of Iowa Press, 1988), chap. 4.

6. Mumford, *Technics and Civilization*, 361–362.

7. Trouillot, *Silencing the Past*, chap. 3.

8. Jackson, "Rethinking Repair," 233.

9. Appadurai, and Alexander, *Failure*, 13.

10. Appadurai, and Alexander, *Failure*, 13.

11. Appadurai, and Alexander, *Failure*, 13, 19.

12. Stuart Hall, "The Toad in the Garden: Thatcherism Among the Theorists," in *Marxism and the Interpretation of Culture*, ed. Cary Nelson and Lawrence Grossberg (Urbana: University of Illinois Press, 1988), 44; and Lorraine Daston, "Introduction: The Hidden History of Rules," in *Rules: A Short History of What We Live By* (Princeton, NJ: Princeton University Press, 2022), 10.

Bibliography

Abbate, Janet. *Inventing the Internet.* Cambridge, MA: MIT Press, 2000.

Abbate, Janet. *Recoding Gender: Women's Changing Participation in Computing.* Cambridge, MA: MIT Press, 2012.

Abraham, Tara. *Rebel Genius: Warren S. McCulloch's Transdisciplinary Life in Science.* Cambridge, MA: MIT Press, 2016.

Adas, Michael. *Machines as the Measure of Men: Science, Technology, and Ideologies of Western Dominance.* Ithaca, NY: Cornell University Press, 2015.

Ahmed, Sara. "Affective Economies." *Social Text* 22, no. 2 (Summer 2004): 117–139.

Ahmed, Sara. *Queer Phenomenology: Orientations, Objects, Others.* Durham, NC: Duke University Press, 2006.

Ahmed, Sara. *What's the Use?: On the Uses of Use.* Durham, NC: Duke University Press Books, 2019.

Aitken, Hugh G. J. *Scientific Management in Action: Taylorism at Watertown Arsenal, 1908–1915.* Princeton, NJ: Princeton University Press, 1985.

Alborn, Timothy L. *Conceiving Companies: Joint-Stock Politics in Victorian England.* New York: Routledge, 1998.

Alder, Ken. "America's Two Gadgets: Of Bombs and Polygraphs." *Isis* 98, no. 1 (2007): 124–137.

Alder, Ken. *Engineering the Revolution: Arms and Enlightenment in France, 1763–1815.* Princeton, NJ: Princeton University Press, 1997.

Alder, Ken. "Introduction." *Isis* 98, no. 1 (2007): 80–83.

Alder, Ken. "Making Things the Same: Representation, Tolerance and the End of the Ancien Régime in France." *Social Studies of Science* 28, no. 4 (1998): 499–545.

Alexander, Jennifer Karns. *The Mantra of Efficiency: From Waterwheel to Social Control.* Baltimore, MD: Johns Hopkins University Press, 2008.

Alexander, Neta. "Rage Against the Machine: Buffering, Noise, and Perpetual Anxiety in the Age of Connected Viewing." *Cinema Journal* 56, no. 2 (2017): 1–24.

Alford, Leon Pratt. *Henry Laurence Gantt: Leader in Industry*. New York: American Society of Mechanical Engineers, 1934.

Allen, Michael Thad, and Gabrielle Hecht, eds. *Technologies of Power: Essays in Honor of Thomas Parke Hughes and Agatha Chipley Hughes*. Cambridge, MA: MIT Press, 2001.

Anderson, Warwick. *The Cultivation of Whiteness: Science, Health and Racial Destiny in Australia*. Carlton, Australia: Melbourne University Press, 2002.

Appadurai, Arjun, and Neta Alexander. *Failure*. New York: Polity, 2019.

Arasse, Daniel. *The Guillotine and the Terror*. New York: Penguin, 1989.

"Arts." *Journal de Paris* 39 (February 8, 1787).

"Arts." *Journal de Paris* 41 (February 10, 1789).

Atanasoski, Neda, and Kalindi Vora. *Surrogate Humanity*. Durham, NC: Duke University Press, 2019.

Athanassoglou-Kallmyer, Nina. "Gericault's Severed Heads and Limbs: The Politics and Aesthetics of the Scaffold." *The Art Bulletin* 74, no. 4 (1992): 599–618.

Aufhauser, R. Keith. "Slavery and Scientific Management." *Journal of Economic History* 33, no. 4 (1973): 811–824.

Baasner, Frank. *Der Begriff "sensibilité" im 18. Jahrhundert: Aufstieg und Niedergang eines Ideals*. Heidelberg: C. Winter, 1988.

Babintseva, Ekaterina Igorevna. "Cyberdreams of the Information Age: Learning with Machines in the Cold War United States and the Soviet Union." PhD diss., University of Pennsylvania, 2020.

Bainbridge-Bell, L. H. "The Need for Clearer Circuit Diagrams." *Engineering* 170 (1950): 260.

Banta, Martha. *Taylored Lives: Narrative Productions in the Age of Taylor, Veblen, and Ford*. Chicago: University of Chicago Press, 1993.

Baptist, Edward E. "Toward a Political Economy of Slave Labor: Hands, Whipping-Machines, and Modern Power." In *Slavery's Capitalism*, edited by Sven Beckert and Seth Rockman. Philadelphia: University of Pennsylvania Press, 2016.

Barad, Karen. *Meeting the Universe Halfway: Quantum Physics and the Entanglement of Matter and Meaning*. Durham, NC: Duke University Press, 2007.

Barak, On. *On Time: Technology and Temporality in Modern Egypt*. Berkeley: University of California Press, 2013.

Baran, Paul. *Reliable Digital Communications Systems Using Unreliable Network Repeater Nodes*. Santa Monica, CA: RAND Corporation, 1960.

Barrett, James R. "Americanization from the Bottom Up: Immigration and the Remaking of the Working Class in the United States, 1880–1930." *Journal of American History* 79, no. 3 (1992): 996–1020.

Bayly, C. A. *The Birth of the Modern World, 1780–1914*. London: Wiley-Blackwell, 2003.

Beccaria, Cesare. *Dei Delitti e Delle Pene*. Rome: Newton Compton, 2012.

Beccaria, Cesare. *Traité des délits et des peines*. Paris: Hélie, 1856.

Beck, Ulrich. *Risk Society: Towards a New Modernity*. Translated by Mark Ritter. Thousand Oaks, CA: Sage, 1992.

Beckert, Sven. "Slavery Takes Command." In *Empire of Cotton: A Global History*. New York: Knopf, 2014.

Behrent, Michael C. "Foucault and Technology." *History and Technology* 29, no. 1 (2013): 54–104.

Bell, Daniel. *The End of Ideology: On the Exhaustion of Political Ideas in the Fifties*. New York: Free Press, 1960.

Bell, David A. "Little Spartas: What Causes Some Cities to Become Sites of Revolution?" *Nation*, January 25, 2018.

Bender, Daniel E. *American Abyss: Savagery and Civilization in the Age of Industry*. Ithaca, NY: Cornell University Press, 2013.

Benjamin, Ruha. *Race After Technology: Abolitionist Tools for the New Jim Code*. New York: Polity, 2019.

Benjamin, Walter. *Das Passagen-Werk. Gesammelte Schrifte*. Vol. 5:2 [Arcades Project]. Frankfurt: Suhrkamp, 1982.

Bennett, Jane. *Vibrant Matter*. Durham, NC: Duke University Press, 2009.

Berman, Marshall. *All That Is Solid Melts into Air: The Experience of Modernity*. New York: Penguin, 1988.

Bertucci, Paola. *Artisanal Enlightenment: Science and the Mechanical Arts in Old Regime France*. New Haven, CT: Yale University Press, 2017.

Bibb, Henry. "Narrative of the Life and Adventures of Henry Bibb, An American Slave." In *Puttin' on Ole Massa: The Slave Narratives of Henry Bibb, William Wells Brown, and Solomon Northop*, edited by Gilbert Osofsky. New York: Harper and Row, 1969.

Bibb, Henry. *Narrative of the Life and Adventures of Henry Bibb, an American Slave, Written by Himself*. Chapel Hill: University of North Carolina at Chapel Hill Library, 2018.

Bijker, Wiebe E., and John Law, eds. *Shaping Technology / Building Society: Studies in Sociotechnical Change*. Cambridge, MA: MIT Press, 1992.

Bijker, Wiebe E., Thomas P. Hughes, and Trevor Pinch, eds. *The Social Construction of Technological Systems: New Directions in the Sociology and History of Technology*. Cambridge, MA: MIT Press, 2012.

Boddice, Rob, and Mark Smith. *Emotion, Sense, Experience*. New York: Cambridge University Press, 2020.

Borg, Kevin L. *Auto Mechanics: Technology and Expertise in Twentieth-Century America*. Baltimore, MD: Johns Hopkins University Press, 2007.

Braithwaite, Frederick. "On the Fatigue and Consequent Fracture of Metals." *Minutes of the Proceedings of the Institution of Civil Engineers* 13 (1854): 463–467.

Brandeis, Louis B. "Scientific Management and the Railroads." *Engineering Magazine*, 1912.

Brattain, Walter H. "Genesis of the Transistor." *The Physics Teacher* 6, no. 3 (1968): 109–114.

Braun, Lundy. *Breathing Race into the Machine: The Surprising Career of the Spirometer from Plantation to Genetics*. Minneapolis: University of Minnesota Press, 2014.

Bray, Francesca. "Gender and Technology." *Annual Review of Anthropology* 36, no. 1 (2007): 37–53.

Bray, Francesca. *Technology and Gender: Fabrics of Power in Late Imperial China*. Berkeley: University of California Press, 1997.

Bray, Francesca. *Technology, Gender and History in Imperial China: Great Transformations Reconsidered*. New York: Routledge, 2013.

Brinkley, Alan. "The Illusion of Unity in Cold War Culture." In *Rethinking Cold War Culture*, edited by Peter J. Kuznick and James Gilbert. Washington, DC: Smithsonian Institution Press, 2001.

Brock, André, Jr. *Distributed Blackness: African American Cybercultures*. New York: New York University Press, 2020.

Brock, David C. "From Automation to Silicon Valley: The Automation Movement of the 1950s, Arnold Beckman, and William Shockley." *History and Technology* 28, no. 4 (2012): 375–401.

Bronstein, Jamie L. *Caught in the Machinery: Workplace Accidents and Injured Workers in Nineteenth-Century Britain*. Stanford, CA: Stanford University Press, 2008.

Broussard, Meredith. *Artificial Unintelligence: How Computers Misunderstand the World*. Cambridge, MA: MIT Press, 2018.

Brown, Bill. "Thing Theory." *Critical Inquiry* 28 (2001): 1–22.

Brown, E. H. *The Corporate Eye: Photography and the Rationalization of American Commercial Culture, 1884–1929*. Baltimore, MD: Johns Hopkins University Press, 2008.

Brown, Philip H. "The Negro Migrant." *Bulletin of the U.S. Bureau of Labor Statistics*, no. 355 (March 1924): 47–52.

Bryan, Glenn L. "Electronics Trouble Shooting: A Behavioral Analysis," Technical Report No. 13. Department of Psychology, University of Southern California, LA, March 1956.

Bryan, Glenn L. "Training of Electronic Maintenance Technicians." *Naval Research Reviews*, 1960–11: 9–15.

Buchaca, Nicolai, and John Brinda Jr. "Maintainability Simulation and Demonstration Equipment Description and Application." *IEEE Transactions on Product Engineering and Production* 7, no. 1 (1963): 6–12.

Buchanan, Thomas C. *Black Life on the Mississippi: Slaves, Free Blacks, and the Western Steamboat World*. Chapel Hill: University of North Carolina Press, 2004.

Buchanan, Thomas C. "Rascals on the Antebellum Mississippi: African American Steamboat Workers and the St. Louis Hanging of 1841." *Journal of Social History* 34, no. 4 (2001): 797–816.

Burke, John G. "Bursting Boilers and the Federal Power." *Technology and Culture* 7, no. 1 (1966): 1–23.

Burnham, John. *Accident Prone: A History of Technology, Psychology, and Misfits of the Machine Age*. Chicago: University of Chicago Press, 2009.

Burrell, Jenna. *Invisible Users: Youth in the Internet Cafés of Urban Ghana*. Cambridge, MA: MIT Press, 2012.

Buss, Sarah. "Appearing Respectful: The Moral Significance of Manners." *Ethics* 109, no. 4 (1999).

Cabanis, Pierre Jean Georges. "Note sur l'opinion de MM. Olsmer et Sömmering, et du citoyen Sue, touchant le supplice de la guillotine." In *Œuvres philosophiques*. Paris: Presses universitaires de France, 1954.

Calendrier Musical Universel. Paris: Prault, 1788.

Calhoun, William P. *The Caucasian and the Negro in the United States*. Columbia, SC: R. L. Bryan, 1902.

Cameron, Laura. "Histories of Disturbance." *Radical History Review* no. 74 (1999): 5–24.

Camp, Stephanie. *Closer to Freedom: Enslaved Women and Everyday Resistance in the Plantation South*. Chapel Hill: University of North Carolina Press, 2004.

Campt, Tina M. *Listening to Images*. Durham, NC: Duke University Press, 2017.

Canguilhem, Georges. "Le Cerveau et la pensée," in *Georges Canguilhem: Philosophe, historien des sciences* (Paris: Albin Michel, 1993), 11–33.

Carby, Hazel V. *Imperial Intimacies: A Tale of Two Islands*. New York: Verso, 2019.

Carney, Judith A. *Black Rice: The African Origins of Rice Cultivation in the Americas*. Cambridge, MA: Harvard University Press, 2009.

Carol, Anne. *Physiologie de la veuve: Une histoire médicale de la guillotine*. Seyssel, France: Champ Vallon, 2012.

Carrabine, Eamonn. "The guillotine: Shadow, spectacle and the terror." *Crime, Media, Culture* 20, no. 4 (2023): 349-369.

Cartwright, Robert. Comment on Frederick W. Taylor's "Notes on Belting." *Transactions of the American Society of Mechanical Engineers*, 15 (1893).

Cartwright, Samuel A. "Report on the Diseases and Physical Peculiarities of the Negro Race." *New Orleans Medical and Surgical Journal*, VII (May 1851): 691–715.

Chakrabarty, Dipesh. *Provincializing Europe: Postcolonial Thought and Historical Difference*. Princeton, NJ: Princeton University Press, 2000.

Chaplin, Joyce. *An Anxious Pursuit: Agricultural Innovation and Modernity in the Lower South, 1730–1815*. Chapel Hill: University of North Carolina Press, 1993.

Charavay, Étienne. *Revue des documents historiques: Suite de pièces curieuses et inédites, publiées avec des notes et des commentaires*. Paris: A. Lemerre, 1876.

Chereau, Achille. *Guillotin et la guillotine*. Paris: Bureaux de l'Union médicale, 1870.

Chinard, Gilbert, ed. *L'apothéose de Benjamin Franklin*. Paris: Librairie Orientale and Américaine, 1955.

Chisick, Harvey. *The Limits of Reform in the Enlightenment: Attitudes Toward the Education of the Lower Classes in Eighteenth-Century France*. Princeton, NJ: Princeton University Press, 2014.

Chronique de Paris. April 16, 1792 (no. 118).

Chun, Wendy Hui Kyong. "Introduction: Race and/as Technology; or, How to Do Things to Race." *Camera Obscura (Durham, NC)* 24, no. 1 (2009): 7–35.

Cipolla, Carlo M. *Clocks and Culture, 1300–1700*. New York: W. W. Norton & Company, 2003.

Claretie, Jules. *Camille Desmoulins and His Wife: Passages from the History of the Dantonists*. Translated by Frances Cashel Hoey. London: Smith, Elder and Co., 1876.

Clark, Sue Ainslie, and Edith Wyatt. *Making Both Ends Meet: The Income and Outlay of New York Working Girls*. New York: Macmillan Company, 1911.

Clark, Wallace. *The Gantt Chart*. New York: Ronald Press, 1922.

Clouzot, Henri. *Les Meubles du VIIIe siècle*. Paris: Albert Monacé, 1780.

Coclanis, Peter A. "How the Low Country Was Taken to Task: Slave-Labor Organization in Coastal South Carolina and Georgia." In *Slavery, Secession, and Southern History*, edited by Robert Louis Paquette and Louis A. Ferleger. Charlottesville: University of Virginia Press, 2000.

Coen, Deborah R. *The Earthquake Observers: Disaster Science from Lisbon to Richter*. Chicago: University of Chicago Press, 2012.

Cohen, Adam Max. *Technology and the Early Modern Self*. New York: Palgrave Macmillan, 2009.

Cohen-Cole, J. "Reason Beyond RAND: Did Enlightenment Values Persist Among Cold War Intellectuals?" *History of the Human Sciences* 31 (2017): 122–126.

Cohen-Cole, Jamie. *The Open Mind: Cold War Politics and the Sciences of Human Nature*. Chicago: University of Chicago Press, 2016.

Collins, H. M. "The Tea Set: Tacit Knowledge and Scientific Networks." *Science Studies* 4, no. 2 (1974): 165–185.

Cook, James W. "Dancing Across the Color Line: A Story of Mixtures and Markets in New York's Five Points." *Commonplace* 4, no. 1 (2003).

Cook, James W. "Lane, William Henry [*performing name* Juba, Master Juba] (c. 1824–c. 1854), Dancer." *Oxford Dictionary of National Biography* (October 3, 2013).

Cooke, Bill. "The Denial of Slavery in Management Studies." *Journal of Management Studies* 40, no. 8 (2003): 1895–1918.

Cooper, Joel. "Human-Initiated Failures and Malfunction Reporting." *IRE Transactions on Human Factors in Electronics* HFE-2, no. 2 (1961): 107.

Cooter, Roger, and Bill Luckin, eds. *Accidents in History: Injuries, Fatalities, and Social Relations*. Amsterdam: Rodopi, 1997.

Cowan, Ruth Schwartz. *More Work for Mother: The Ironies of Household Technology from the Open Hearth to the Microwave*. New York: Basic Books, 1983.

Crook, Tom. *Governing Systems: Modernity and the Making of Public Health in England, 1830–1910*. Berkeley: University of California Press, 2016.

Crook, Tom, and Mike Esbester, eds. *Governing Risks in Modern Britain: Danger, Safety and Accidents, c. 1800–2000*. London: Palgrave Macmillan, 2016.

Crook, Tom, and Mike Esbester. "Risk and the History of Governing Modern Britain, c. 1800–2000." In *Governing Risks in Modern Britain: Danger, Safety and Accidents, c. 1800–2000*, edited by Tom Crook and Mike Esbester. London: Palgrave Macmillan, 2016.

Cruikshank, Barbara. *The Will to Empower.* Ithaca, NY: Cornell University Press, 1999.

Cutler, Janet. "Neglected Historical Sources: The Railway Inspectors Reports." *History Workshop Journal* 27, no. 1 (1989): 174–175.

Czeh, Robert. "Studies of Complex Behavior and Their Relation to Trouble Shooting in Electronic Equipment." US Army Contract DA-49-105-qm-1. Alexandria, VA: Human Resources Research Organization, 1957.

Daly, Nicholas. "Railway Novels: Sensation Fiction and the Modernization of the Senses." *ELH* 66, no. 2 (1999): 461–487.

Darnton, Robert. *The Literary Underground of the Old Regime.* Cambridge, MA: Harvard University Press, 1982.

Daston, Lorraine. "Introduction: The Hidden History of Rules." In *Rules: A Short History of What We Live By*. Princeton, NJ: Princeton University Press, 2022.

Daston, Lorraine, ed. *Things That Talk. Object Lessons from the History of Art and Science.* New York: Zone Books, 2004.

Daston, Lorraine, and Peter Galison. "The Image of Objectivity." *Representations* 40 (1992): 81–128.

Daston, Lorraine, and Peter Galison. *Objectivity*. New York: Zone Book, 2007.

Davidson, Arnold. "How to Do the History of Psychoanalysis: A Reading of Freud's Three Essays on the Theory of Sexuality." *Critical Inquiry* 13, no. 2 (1987): 252–277.

Davidson, Arnold I. "Spiritual Exercises and Ancient Philosophy: An Introduction to Pierre Hadot." *Critical Inquiry* 16, no. 3 (1990): 475–482.

De Baecque, Antoine. *The Body Politic: Corporeal Metaphor in Revolutionary France, 1770–1800.* Stanford, CA: Stanford University Press, 1997.

de la Peña, Carolyn. "The History of Technology, the Resistance of Archives, and the Whiteness of Race." *Technology and Culture* 51, no. 4 (2010): 919–937.

de la Tynna, J. *Almanach du commerce de Paris, des départemens de la France et des principales villes du monde Année 1811*. Paris: de la Tynna, 1811.

de Laet, Marianne, and Annemarie Mol. "The Zimbabwe Bush Pump: Mechanics of a Fluid Technology." *Social Studies of Science* 30, no. 2 (2000): 225–263.

de Vaucanson, Jacques. "Construction de nouveaux moulins à organciner les foyes." *Mercure de France*, June 1751.

de Voe, Thomas F. *The Market Book, Containing a Historical Account of the Public Markets in the Cities of New York, Boston, Philadelphia and Brooklyn, with a Brief Description of Every Article of Human Food Sold Therein, the Introduction of Cattle in America, and Notices of Many Remarkable Specimens. Vol. 1.* New York: Printed for the author, 1862.

Denby, David J. *Sentimental Narrative and the Social Order in France, 1760–1820*. New York: Cambridge University Press, 1994.

Denis, David J. "Why Do Maintenance and Repair Matter." In *The Routledge Companion to Actor-Network Theory*, edited by Anders Blok, Ignacio Farias, and Celia Roberts. New York: Routledge, 2019.

Denis, Jérôme, and David Pontille. "Beyond Breakdown: Exploring Regimes of Maintenance." *Continent* 6, no. 1 (2017): 13–17.

Derksen, Maarten. "Turning Men into Machines? Scientific Management, Industrial Psychology, and the 'Human Factor.'" *Journal of the History of the Behavioral Sciences* 50, no. 2 (2014): 148–165.

Dewey, John. *How We Think*. Boston: D.C. Heath & Co., 1910.

Awa Hanane Diagne, Emily Grenon, Syeda Hasan, et al., "The Black Androids: A Chrestomathy." *Historical Studies in the Natural Sciences* 54, no. 3 (2024): 407–421.

Dickens, Charles. *American Notes for General Circulation*. London: Chapman and Hall, 1850.

Diderot, Denis. "Lettre sur les aveugles à l'usage de ceux qui voyent" (1749). In *Ouvres philosophiques*, edited by Paul Vernière. Paris: Garnier, 1998.

Diderot, Denis, and Pierre Tarin. "Anatomie." In *Encyclopédie, ou dictionnaire raisonné des sciences, des arts et des métiers, etc.*, edited by Denis Diderot and Jean le Rond d'Alembert, vol. 1. Paris: Le Breton, 1751.

Dinerstein, Joel. *Swinging the Machine: Modernity, Technology, and African American Culture Between the World Wars*. Amherst: University of Massachusetts Press, 2003.

"Discrimination Against Enlightenment and Efficiency." *Afro-American*, 1912.

Disney, Henry W. *The Law of Carriage by Railway*. London: Stevens and Sons, 1905.

Dobbin, Frank. *Forging Industrial Policy: The United States, Britain, and France in the Railway Age*. Cambridge: Cambridge University Press, 1994.

Donaldson, Gary A. "A Window on Slave Culture: Dances at Congo Square in New Orleans, 1800–1862." *Journal of Negro History* 69, no. 2 (1984): 63–72.

Drury, Horace Bookwalter. *Scientific Management: A History and Criticism*. New York: Columbia University Press, 1915.

Drury, Horace B. *Scientific Management: A History and Criticism*. 2nd ed. New York: Columbia University, 1918.

Du Bois, Louis. *Recherches historiques et physiologiques sur la guillotine, et détails sur Sanson.* Paris: France, 1843.

Du Bois, W. E. B. "The Negro Artisan." In *The Negro Artisan: Report of a Social Study Made under the Direction of Atlanta University; Together with the Proceedings of the Seventh Conference for the Study of the Negro Problems, Held at Atlanta University, on May 27th, 1902,* edited by W. E B. Du Bois. Atlanta, GA: Atlanta University Press, 1902.

Du Bois, W. E. B. *The Philadelphia Negro: A Social Study*. Philadelphia: University of Pennsylvania, 1996. Originally published in 1899.

Du Bois, W. E. B. *The Souls of Black Folk.* New York: Dover Publications, 1903.

Dusinberre, William. *Them Dark Days: Slavery in the American Rice Swamps.* New York: Oxford University Press, 1996.

Editorial Correspondence. "Canadian Pacific Shop Management." *American Machinist* 35 (December 21, 1911): 1164–1167.

Edgerton, David. *The Shock of the Old: Technology and Global History Since 1900.* New York: Oxford University Press, 2011.

Edwards, Paul N. *The Closed World: Computers and the Politics of Discourse in Cold War America.* Cambridge, MA: MIT Press, 1997.

Edwards, Paul N., and Gabrielle Hecht. "History and the Technopolitics of Identity: The Case of Apartheid South Africa." *Journal of Southern African Studies* 36, no. 3 (2010): 619–639.

"Efficiency Prize." *Chicago Defender,* May 17, 1913.

Elias, Norbert. *The Civilizing Process: The History of Manners,* vol. 1. Translated by Edmund Jephcott. Oxford: Basil Blackwell, 1978.

Ellison, Ralph. *Invisible Man.* New York: Random House, 1952.

Emerson, Harrington. "Efficiency as a Basis for Operation and Wages: The Efficiency System in Operation." *Engineering Journal* 36 (1909).

Erickson, Paul, Judy L. Klein, Lorraine Daston, Rebecca Lemov, Thomas Sturm, and Michael D. Gordin. *How Reason Almost Lost Its Mind: The Strange Career of Cold War Rationality*. Chicago: University of Chicago Press, 2013.

Esbester, Mike, and J. Wetmore, eds. "(Auto) Mobility, Accidents, and Danger." Special issue of *Technology and Culture* 52 (2015).

Eustace, Nicole, Eugenia Lean, Julie Livingston, Jan Plamper, William M. Reddy, and Barbara H. Rosenwein. "AHR Conversation: The Historical Study of Emotions." *American Historical Review* 117, no. 5 (2012): 1487–1531.

"Evidence Taken by the Interstate Commerce Commission in the Matter of Proposed Advances in Freight Rates by Carriers." August to December 1910, vol. 4, 61st Congress three-day session, Washington, DC, 1911.

"Extrait de registres de l'academie chirurgie, du jeudi 31 décembre 1778." *L'Esprit des Journaux* 8, no 2 (1779): 344–345.

Eze, Emmanuel Chukwudi. *Race and the Enlightenment: A Reader*. Malden, MA: Blackwell, 1997.

Farish, Matthew. "The Lab and the Land: Overcoming the Arctic in Cold War Alaska." *Isis* 104, no. 1 (2013): 1–29.

Farish, Matthew. *The Contours of America's Cold War*. Minneapolis: University of Minnesota Press, 2010.

Farmer, William, ed. "Memoir of William Wells Brown." In *Three Years in Europe: Or, Places I Have Seen and People I Have Met*. New York: Cambridge University Press, 2014.

Faust, Drew Gilpin. *James Henry Hammond and the Old South: A Design for Mastery*. Baton Rouge: Louisiana State University Press, 1985.

Fehér, Ferenc. *The French Revolution and the Birth of Modernity*. Berkeley: University of California Press, 1990.

Ferguson, Charles. *Philadelphia Public Ledger*. December 10, 1916.

Ferguson, Charles. *The Revolution Absolute*. New York: Dodd, Mead, 1918.

Ferguson, Eugene S. *Engineering and the Mind's Eye*. Cambridge, MA: MIT Press, 1994.

Festa, Lynn. *Sentimental Figures of Empire in Eighteenth-Century Britain and France*. Baltimore, MD: Johns Hopkins University Press, 2006.

Figlio, Karl. "What Is an Accident?" In *A Social History of Occupational Health*, edited by Paul Weindling. London: Croom Helm, 1985.

Fleischmann, Hector. *La Guillotine En 1793: D'après Des Documents inédits Des Archives Nationales*. Paris: Librairie des Publications Modernes, 1908.

Fortun, Kim. *Advocacy After Bhopal: Environmentalism, Disaster, New Global Orders*. Chicago: University of Chicago Press, 2001.

Foster, George G. *New York by Gas-Light and Other Urban Sketches*. Berkeley: University of California Press, 1990.

Foucault, Michel. *Discipline and Punish: The Birth of the Prison*. Translated by Alan Sheridan. New York: Vintage, 1995.

Foucault, Michel. *Dits et écrits*, IV: 1980–1988. Paris: NRF Gallimard, 1994.

Foucault, Michel. *The Hermeneutics of the Subject: Lectures at the Collège De France, 1981–1982*. Edited by Frédéric Gros, translated by Graham Burchell. New York: Picador, 2006.

Foucault, Michel. *The History of Sexuality: The Care of the Self.* Translated by Robert Hurley. New York: Vintage, 1979.

Foucault, Michel. *The Order of Things: An Archaeology of the Human Sciences.* New York: Routledge, 2002.

Foucault, Michel. "Space, Knowledge, and Power." In *The Foucault Reader*, edited by Paul Rabinow. London: Penguin, 1986.

Fouché, Rayvon. *Black Inventors in the Age of Segregation: Granville T. Woods, Lewis H. Latimer, and Shelby J. Davidson.* Baltimore, MD: Johns Hopkins University Press, 2003.

Fouché, Rayvon. "Not Made for Black History Month: Lewis Latimer and Technological Assimilation." In *Appropriating Technology: Vernacular Science and Social Power*, edited by Ron Eglash. Minneapolis: University of Minnesota Press, 2004.

Fouché, Rayvon. "Say It Loud, I'm Black and I'm Proud: African Americans, American Artifactual Culture, and Black Vernacular Technological Creativity." *American Quarterly* 58, no. 3 (2006): 639–661.

Fouché, Rayvon. "The Wretched of the Gulf: Racism, Technological Dramas, and Black Politics of Technology." *The Black Scholar* 36, no. 4 (2006): 7–12.

Freedgood, Elaine. *Victorian Writing About Risk: Imagining a Safe England in a Dangerous World.* New York: Cambridge University Press, 2006.

Freeman, Michael J. *Railways and the Victorian Imagination.* New Haven, CT: Yale University Press, 1999.

Friedland, Paul. *Seeing Justice Done: The Age of Spectacular Capital Punishment in France.* Oxford: Oxford University Press, 2012.

Fuentes, Marisa J. *Dispossessed Lives: Enslaved Women, Violence, and the Archive.* Philadelphia: University of Pennsylvania Press, 2016.

Fyfe, Paul. *By Accident or Design: Writing the Victorian Metropolis.* New York: Oxford University Press, 2015.

Fyfe, Paul. "Illustrating the Accident: Railways and the Catastrophic Picturesque in the Illustrated London News." *Victorian Periodicals Review* 46, no. 1 (2013): 61–91.

Galanter, Eugene, ed. *Teaching Machines for Training of Military Personnel in Maintenance of Electronic Equipment.* New York: Wiley, 1959.

Galison, Peter. "An Accident of History." In *Atmospheric Flight in the Twentieth Century*, edited by Peter Galison and Alex Roland. Dordrecht, Netherlands: Springer, 2000.

Galison, Peter. "On the Building, Crashing, and Thinking of Technologies and Self-hood." In *Art in the Anthropocene: Encounters Among Aesthetics, Politics, Environments, and Epistemologies,* edited by Etienne Turpin and Heather Davies. London: Open Humanities Press, 2015.

Galison, Peter. "Ontology of the Enemy: Norbert Wiener and the Cybernetic Vision." *Critical Inquiry* 21 (1994): 228–266.

Galton, Douglas. "Report to the Lords of the Committee of Privy Council for Trade and Foreign Plantations of the Proceedings of the Department Relating to Railways for the Year 1858." Appendix 1. London: HM Stationery Office, 1856.

Galton, Douglas. "Report to the Lords of the Committee of the Privy Council for Trade upon the Accidents Which Have Occurred on Railways During the Year 1858." C Series. London: HM Stationery Office, 1859.

Gandy, Oscar H. "Instructional Technology: The Reselling of the Pentagon." PhD diss., Stanford University, 1976.

Gantt, Henry L. "Democratic Shop Methods." Speech, De Laval Steam Turbine Co., Henry L. Gantt Papers. New York University, January 29, 1919.

Gantt, Henry Laurence. "A Bonus System of Rewarding Labor: Being a System of Task, with Instruction Cards and a Bonus." *Transactions of the American Society of Mechanical Engineers* 23 (1902): 341–372.

Gantt, Henry Laurence. "Efficiency and Idleness." *Industrial Management* 56, no. 5 (1918): 377.

Gantt, Henry Laurence. *Organizing for Work.* New York: Harcourt, Brace and Howe, 1919.

Gantt, Henry Laurence. "Scheduling Locomotive Repair Work on the Canadian Pacific Railway." *Industrial Engineering and the Engineering Digest* 8, no. 5 (November 1910): 380–383.

Gantt, Henry Laurence. "The Religion of Democracy." *Industrial Management* 57, no. 5 (1919): 404–405.

Gantt, Henry Laurence. "Work, Wages, and Profits." *Engineering Magazine,* 1910.

Gantt, Henry Laurence. *Work, Wages and Profits.* 2nd ed. New York: The Engineering Magazine, 1913.

Gay, Peter. *The Enlightenment: An Interpretation,* vol. 2, *The Science of Freedom.* New York: Knopf, 2013.

"General Report to the Board of Trade upon the Accidents Which Have Occurred on the Railways of the United Kingdom During the Year 1881." Excerpt, August 24, 1881. London: HM Stationery Office, 1882.

Gerovitch, Slava. *From Newspeak to Cyberspeak: A History of Soviet Cybernetics*. Cambridge, MA: MIT Press, 2004.

Ghamari-Tabrizi, Sharon. *The Worlds of Herman Kahn: The Intuitive Science of Thermonuclear War*. Cambridge, MA: Harvard University Press, 2005.

Giedion, Sigfried. *Mechanization Takes Command: A Contribution to Anonymous History*. New York: Oxford University Press, 1948.

Gillispie, Charles C. *Science and Polity in France: The End of the Old Regime*. Princeton, NJ: Princeton University Press, 2004.

Gilroy, Paul. *The Black Atlantic: Modernity and Double Consciousness*. Cambridge, MA: Harvard University Press, 1993.

Gioia, Ted. *The History of Jazz*. 3rd ed. New York: Oxford University Press, 2021.

Glausser, Wayne. "Three Approaches to Locke and the Slave Trade." *Journal of the History of Ideas* 51, no. 2 (1990): 199–216.

Golden, John. "'Scientific Management' in the Textile Industry." *American Federationist* 18, no. 8 (1911): 603–604.

Goldstein, Jan. *The Post-Revolutionary Self: Politics and Psyche in France, 1750–1850*. Cambridge, MA: Harvard University Press, 2005.

Gompers, Samuel. "The Miracles of 'Efficiency.'" *American Federationist* 18, no. 4 (1911): 273–279.

Goodman, Dena. *The Republic of Letters: A Cultural History of the French Enlightenment*. Ithaca, NY: Cornell University Press, 1994.

Goodspeed, D. J. *A History of the Defence Research Board of Canada*. Ottawa: Queen's Printer, 1958.

Gorman, Michael E., and W. Bernard Carlson. "Interpreting Invention as a Cognitive Process: The Case of Alexander Graham Bell, Thomas Edison, and the Telephone." *Science, Technology, and Human Values* 15, no. 2 (1990): 131–164.

Govaerts, Sander. *Armies and Ecosystems in Premodern Europe*. Leeds: Arc Humanities Press, 2021.

Grace, Joshua. *African Motors: Technology, Gender, and the History of Development*. Durham, NC: Duke University Press, 2021.

Graham, Laurel D. "Domesticating Efficiency: Lillian Gilbreth's Scientific Management of Homemakers, 1924–1930." *Signs* 24, no. 3 (1999): 633–675.

Graham, Stephen, and Nigel Thrift. "Out of Order: Understanding Repair and Maintenance." *Theory, Culture and Society* 24, no. 3 (2007): 1–25.

Green, Judith. *Risk and Misfortune: A Social Construction of Accidents*. London: UCL Press, 1997.

Greenblatt, Stephen. *Marvelous Possessions: The Wonder of the New World*. New York: Oxford University Press, 1992.

Greenblatt, Stephen. *Renaissance Self-Fashioning: From More to Shakespeare*. Chicago: University of Chicago Press, 2005.

Greenhow, Conrad H. *An Exposition of the Danger and Deficiencies of the Present Mode of Railway Construction: With Suggestions for Its Improvement*. London: John Weale, 1846.

Gregory, James. *The Southern Diaspora: How the Great Migrations of Black and White Southerners Transformed America*. Chapel Hill: University of North Carolina Press, 2005.

Gregory, Mary Efrosini. *Freedom in French Enlightenment Thought*. New York: Peter Lang, 2010.

Griffith, R. Marie. "Apostles of Abstinence: Fasting and Masculinity During the Progressive Era." *American Quarterly* 52, no. 4 (2000): 599–638.

Grollier de Servière, Nicolas. *Recueil d'ouvrages curieux de mathématique et de mécanique, ou Description du cabinet de M. Grollier de Servière*. Lyon: D. Forey, 1719.

Gutman, Herbert G. "Work, Culture, and Society in Industrializing America, 1815–1919." *American Historical Review* 78, no. 3 (1973): 531–588.

Haber, Samuel. *Efficiency and Uplift: Scientific Management in the Progressive Era, 1890–1920*. Chicago: University of Chicago Press, 1964.

Habermas, Jürgen. *The Structural Transformation of the Public Sphere: An Inquiry into a Category of Bourgeois Society*. Translated by Thomas Burger. Cambridge, MA: MIT Press, 1989.

Hacking, Ian. *Rewriting the Soul: Multiple Personality and the Sciences of Memory*. Princeton NJ: Princeton University Press, 2003.

Hacking, Ian. *The Taming of Chance*. New York: Cambridge University Press, 1990.

Hadley, Elaine. *Living Liberalism: Practical Citizenship in Mid-Victorian Britain*. Chicago: University of Chicago Press, 2010.

Hadot, Pierre. *Philosophy as a Way of Life: Spiritual Exercises from Socrates to Foucault*. Edited by Arnold I. Davidson. Malden, MA: Blackwell, 1995.

Hadot, Pierre, Arnold I. Davidson, and Paula Wissing. "Forms of Life and Forms of Discourse in Ancient Philosophy," translated by Arnold I. Davidson and Paula Wissing. *Critical Inquiry* 16, no. 3 (1990): 483–505.

Hadot, Pierre, Jeannie Carlier, and Arnold I. Davidson. *La Philosophie Comme Manière De Vivre*. Paris: A. Michel, 2001.

Halberstam, Jack. *The Queer Art of Failure*. Durham, NC: Duke University Press, 2011.

Hall, Stuart. "The Toad in the Garden: Thatcherism Among the Theorists." In *Marxism and the Interpretation of Culture*, edited by Cary Nelson and Lawrence Grossberg. Urbana: University of Illinois Press, 1988.

Halsey, Frederick A. "The Premium Plan of Paying for Labor." *Transactions of the American Society of Mechanical Engineers* 22 (1891): 755–780.

Hammonds, Evelyn. "New Technologies of Race." In *Processed Lives: Gender and Technology in Everyday Life*, edited by Jennifer Terry and Melodie Calvert. New York: Routledge, 1997.

Haraway, Donna. "A Cyborg Manifesto: Science, Technology, and Socialist-Feminism in the Late Twentieth Century." In *Simians, Cyborgs, and Women: The Reinvention of Nature*. New York: Routledge, 1990.

Harrington, Ralph. "Railway Safety and Railway Slaughter: Railway Accidents, Government and Public in Victorian Britain." *Journal of Victorian Culture* 8 (2003): 187–207.

Harrington, Ralph. "The Railway Accident: Trains, Trauma, and Technological Crisis in Nineteenth-Century Britain." In *Traumatic Pasts: History, Psychiatry, and Trauma in the Modern Age, 1870–1930*, edited by Mark S. Micale and Paul Lerner. New York: Cambridge University Press, 2010.

Hartman, Saidiya. *Lose Your Mother: A Journey Along the Atlantic Slave Trade Route*. New York: Farrar, Straus and Giroux, 2007.

Hartman, Saidiya. *Wayward Lives, Beautiful Experiments: Intimate Histories of Riotous Black Girls, Troublesome Women, and Queer Radicals*. New York: Norton & Company, 2019.

Hayles, N. Katherine. *How We Became Posthuman: Virtual Bodies in Cybernetics, Literature, and Informatics*. Chicago: University of Chicago Press, 1999.

Haynes, George E. *The Negro at Work During the World War and During Reconstruction: Statistics, Problems, and Policies Relating to the Greater Inclusion of Negro Wage Earners in American Industry and Agriculture*. Washington, DC: Government Printing Office, 1921.

Hays, Samuel P. *Conservation and the Gospel of Efficiency: The Progressive Conservation Movement, 1890–1920*. Pittsburgh, PA: University of Pittsburgh Press, 1999.

Hays, Samuel P. "Woodman, Spare the Tree." In *Conservation and the Gospel of Efficiency: The Progressive Conservation Movement, 1890–1920*. Pittsburgh, PA: University of Pittsburgh Press, 1999.

Headrick, Daniel R. *The Tools of Empire: Technology and European Imperialism in the Nineteenth Century*. New York: Oxford University Press, 1981.

"Hearings before the Special Committee of the House of Representatives to Investigate the Taylor and Other Systems of Shop Management." Washington, DC, 1912.

Hecht, Gabrielle. "Enacting Cultural Identity: Risk and Ritual in the French Nuclear Workplace." *Journal of Contemporary History* 32, no. 4 (1997): 483–507.

Hecht, Gabrielle. *Residual Governance: How South Africa Foretells Planetary Futures.* Durham, NC: Duke University Press 2023.

Hecht, Gabrielle. *The Radiance of France: Nuclear Power and National Identity After World War II.* Cambridge, MA: MIT Press, 2009.

Heidegger, Martin. *Sein Und Seit.* Tübingen, Germany: Max Niemeyer Verlag, 1967.

Heidegger, Martin. *The Question Concerning Technology, and Other Essays.* Translated by William Lovitt. New York: Harper Torchbooks, 1977.

Heims, Steve J. *Constructing a Social Science for Postwar America: The Cybernetics Group, 1946–1953.* Cambridge, MA: MIT Press, 1993.

Helvétius, Claude Adrien. "Letter to President Montesquieu." In *Correspondance Générale D'Helvétius*, vol. 5, Appendicies et Index, edited by Alan Dainard, David Smith, Jean Orsoni, Jonas Steffen, Marie-Thérèse Inguenaud, and Peter Allan. Toronto: University of Toronto Press, 2004.

Henke, Christopher. "The Mechanics of Workplace Order: Toward a Sociology of Repair." *Berkeley Journal of Sociology* 44 (2000): 55–81.

Henney, Keith, ed. *Reliability Factors for Ground Electronic Equipment.* New York: McGraw-Hill, 1956.

Herf, Jeffrey. *Reactionary Modernism: Technology, Culture, and Politics in Weimar and the Third Reich.* Cambridge: Cambridge University Press, 1984.

Hersch, Charles B. *Subversive Sounds: Race and the Birth of Jazz in New Orleans.* Chicago: University of Chicago Press, 2008.

Hersch, Matthew H. *Dark Star: A New History of the Space Shuttle.* Cambridge, MA: MIT Press, 2023.

Heyck, Hunter. *Age of System: Understanding the Development of Modern Social Science.* Baltimore, MD: Johns Hopkins University Press, 2015.

Hicks, Mar. *Programmed Inequality: How Britain Discarded Women Technologists and Lost Its Edge in Computing.* Cambridge, MA: MIT Press, 2017.

Hoffman, Frederick. "Vital Statistics of the Negro." *Arena*, no. 29 (April 1892).

Holloway, Joseph E. *Africanisms in American Culture.* 2nd ed. Bloomington: Indiana University Press, 2005.

Holm, Christiane. "Sentimental Cuts: Eighteenth-Century Mourning Jewelry with Hair." *Eighteenth-Century Studies* 38, no. 1 (2004): 139–143.

Horn, Jeff. "The Other 'Great Fear': Labor Relations, Industrialization, and Revolution." In *The Path Not Taken: French Industrialization in the Age of Revolution, 1750–1830*. Cambridge, MA: MIT Press, 2008.

Hudson, Janet G. *Entangled by White Supremacy: Reform in World War I–Era South Carolina*. Lexington: University Press of Kentucky, 2009.

Hughes, Agatha C., and Thomas P. Hughes, eds. *Systems, Experts, and Computers: The Systems Approach in Management and Engineering, World War II and After*. Cambridge, MA: MIT Press, 2000.

Hughes, Alex. "*The History of Tears: Sensibility and Sentimentality in France*, by Anne Vincent-Buffault." *French History* 6, no. 3 (1992): 381–382.

Hughes, Thomas P. *Networks of Power: Electrification in Western Society, 1880–1930*. Baltimore, MD: Johns Hopkins University Press, 1983.

Hughes, Thomas P. "The Evolution of Large Technological Systems." In *The Social Construction of Technological Systems: New Directions in the Sociology and History of Technology*, edited by Wiebe E. Bijker, Thomas P. Hughes, and Trevor Pinch. Cambridge, MA: MIT Press, 2012.

Igo, Sarah E. *The Averaged American: Surveys, Citizens, and the Making of a Mass Public*. Cambridge, MA: Harvard University Press, 2007.

Immerwahr, Daniel. "All That is Solid Bursts into Flame: Capitalism and Fire in the Nineteenth-Century United States." *Past and Present* 265, no. 1 (2024): 97–138.

Irani, Lilly. "The Cultural Work of Microwork." *New Media & Society* 17, no. 5 (2013): 720–739.

Iron, October 10, 1879.

Israel, Jonathan. *A Revolution of the Mind: Radical Enlightenment and the Intellectual Origins of Modern Democracy*. Princeton, NJ: Princeton University Press, 2009.

Jackson, Steven J. "Rethinking Repair." In *Media Technologies: Essays on Communication, Materiality, and Society*, edited by Tarleton Gillespie, Pablo J. Boczkowski, and Kirsten A. Foot. Cambridge, MA: MIT Press, 2014.

Janes, Regina. "Beheadings." *Representations* 35, no. 1 (1991): 21–51.

J. B. W. "Whittemore's Cotton Gin." *Southern Agriculturist* 8, no. 9 (September 1835): 473–480.

Jenkins, Destin, and Justin Leroy, eds. *Histories of Racial Capitalism*. New York: Columbia University Press, 2021.

Jennings, Humphrey. *Pandaemonium 1660–1886: The Coming of the Machine as Seen by Contemporary Observers*. London: Icon, 2012.

Johnson, Samuel. *A Dictionary of the English Language*. London: W. Strahan, 1755.

Jones, Jacqueline. "'Lifework' and Its Limits: The Problem of Labor in *The Philadelphia Negro*." In *W. E. B. Du Bois, Race, and the City: The Philadelphia Negro and Its Legacy*, edited by Michael B. Katz and Thomas J. Sugrue. Philadelphia: University of Pennsylvania Press, 1998.

Jones-Imhotep, Edward. "Icons and Electronics." *Historical Studies in the Natural Sciences* 38, no. 3 (2008): 405–50.

Jones-Imhotep, Edward. "Paris-Montreal-Babylon: The Modernist Genealogies of Gerald Bull." In *Made Modern: Science and Technology in Canadian History*, edited by Edward Jones-Imhotep and Tina Adcock. Vancouver: UBC Press, 2018.

Jones-Imhotep, Edward. "The Ghost Factories: Histories of Automata and Artificial Life." *History and Technology* 36, no. 1 (2020): 3–29.

Jones-Imhotep, Edward. "The Unfailing Machine: Mechanical Arts, Sentimental Publics and the Guillotine in Revolutionary France." *History of the Human Sciences* 30, no. 4 (2017): 11–31.

Jones-Imhotep, Edward. *The Unreliable Nation: Hostile Nature and Technological Failure in the Cold War*. Cambridge, MA: MIT Press, 2017.

Jones-Imhotep, Edward, and Alexander Offord. "Black Steam: Patents, Portals, and the Counter-Histories of the Victorian Android." In *Victorian Automata: Mechanism and Agency in the Nineteenth Century*, edited by Suzy Anger and Thomas Vranken. New York: Cambridge University Press, 2024.

Jordanova, Ludmilla. "Medical Mediations: Mind, Body and the Guillotine." *History Workshop* 28, no. 1 (1989): 39–52.

Joyce, Patrick. *The Rule of Freedom: Liberalism and the Modern City*. New York: Verso, 2003.

Joyce, Patrick. *The State of Freedom: A Social History of the British State Since 1800*. Cambridge: Cambridge University Press, 2013.

Joyner, Charles. *Down by the Riverside: A South Carolina Slave Community*. Urbana: University of Illinois Press, 1984.

Kaiser, David. "Cold War Requisitions, Scientific Manpower, and the Production of American Physicists After World War II." *Historical Studies in the Physical and Biological Sciences* 33, no. 1 (2002): 131–159.

Kaiser, David. *Drawing Theories Apart: The Dispersion of Feynman Diagrams in Postwar Physics*. Chicago: University of Chicago Press, 2005.

Kaiser, David, Kenji Ito, and Karl Hall. "Spreading the Tools of Theory: Feynman Diagrams in the USA, Japan, and the Soviet Union." *Social Studies of Science* 34, no. 6 (2004): 879–922.

Kanigel, Robert. *The One Best Way: Frederick Winslow Taylor and the Enigma of Efficiency*. New York: Viking, 1997.

Kaplan, Fred. *The Wizards of Armageddon*. Stanford, CA: Stanford University Press, 1991.

Karafantis, Layne, and Daniel Volmar. "Idiot-Proofing the Air Force: Human Factors Engineering and the Crisis of Systems of Maintenance." Paper presented at *The Maintainers II: Labor, Technology, and Social Orders*, April 6–9, 2017, Hoboken, NJ.

Kelley, Robin D. G. *Race Rebels: Culture, Politics, and the Black Working Class*. New York: Free Press, 1996.

Kelley, Robin D. G. "'We Are Not What We Seem': Rethinking Black Working-Class Opposition in the Jim Crow South." *Journal of American History* 80, no. 1 (1993): 75–112.

Kennan, George F. "America's Administrative Response to Its World Problems." In *The American Style: Essays in Value and Performance*, edited by Elting E. Morison. New York: Harper and Brothers, 1957.

Kern, Stephen. *The Culture of Time and Space, 1880–1918*. Cambridge, MA: Harvard University Press, 2003.

Kershaw, Alister. *A History of the Guillotine*. London: John Calder, 1958.

Kingsford, P. W. *Victorian Railwaymen*. London: Frank Cass and Co. Ltd., 1970.

Klein, Naomi. *The Shock Doctrine: The Rise of Disaster Capitalism*. Toronto: Knopf Canada, 2007.

Klein, Ursula. *Experiments, Models, Paper Tools: Cultures of Organic Chemistry in the Nineteenth Century*. Stanford, CA: Stanford University Press, 2003.

Kline, Ronald L. *The Cybernetics Moment; or, Why We Call Our Age the Information Age*. Baltimore, MD: Johns Hopkins University Press, 2015.

Kling, David W. "English Protestantism (1520–1700)." In *A History of Christian Conversion*. New York: Oxford University Press, 2020.

Kmen, Henry A. *Music in New Orleans: The Formative Years, 1791–1841*. Baton Rouge: Louisiana State University Press, 1966.

Kmen, Henry A. "The Roots of Jazz and the Dance in Place Congo: A Re-Appraisal." *Anuario Interamericano de Investigación Musical* 8 (1972): 5–16.

Knowles, Scott Gabriel. *The Disaster Experts: Mastering Risk in Modern America*. Philadelphia: University of Pennsylvania Press, 2013.

Kostal, R. W. *Law and English Railway Capitalism, 1825–1875.* Oxford: Clarendon, 1994.

Krajewski, Markus. *Paper Machines: About Cards and Catalogs, 1548–1929.* Translated by Peter Krapp. Cambridge, MA: MIT Press, 2011.

Krebs, Stefan, and Heike Weber, eds. *The Persistence of Technology: Histories of Repair, Reuse and Disposal.* Bielefeld, Germany: Transcript Verlag, 2021.

Lakwete, Angela. *Inventing the Cotton Gin: Machine and Myth in Antebellum America.* Baltimore, MD: Johns Hopkins University Press, 2005.

Lamb, H. H. *Climate, History, and the Modern World.* 2nd ed. New York: Routledge, 1995.

Landes, Joan B. *Women and the Public Sphere in the Age of the French Revolution.* Ithaca, NY: Cornell University Press, 1988.

Larkin, Brian. *Signal and Noise: Media, Infrastructure, and Urban Culture in Nigeria.* Durham, NC: Duke University Press, 2008.

Larsen, Nella. *Quicksand.* New York: Alfred A. Knopf, 1928.

Latour, Bruno. "Drawing Things Together." In *Representation in Scientific Practice,* edited by Michael Lynch and Steve Woolgar. Cambridge, MA: MIT Press, 1990.

Latour, Bruno. "Give Me a Laboratory and I Will Raise the World." In *Science Observed: Perspectives on the Social Study of Science,* edited by Karin Knorr-Cetina and Michael Mulkay. London: Sage, 1983.

Latour, Bruno. *Reassembling the Social: An Introduction to Actor-Network-Theory.* New York: Oxford University Press, 2005.

Latour, Bruno. "Technology Is Society Made Durable." *Sociology Review* 38, no. 1 (1990): 103–131.

Latour, Bruno. "Visualization and Cognition." In *Knowledge and Society: Studies in the Sociology of Culture Past and Present,* edited by H. Kuklick. Greenwich, CT: JAI Press, 1986.

Latour, Bruno. *Aramis, or, The Love of Technology.* Cambridge, Mass: Harvard University Press, 1996.

Latour, Bruno. *We Have Never Been Modern.* Cambridge, MA: Harvard University Press, 1993.

Latour, Bruno. "When Things Strike Back: A Possible Contribution of 'Science Studies' to the Social Sciences." *British Journal of Sociology* 51 (2020): 107–123.

Latour, Bruno. "Where Are the Missing Masses? The Sociology of Mundane Artifacts." In *Shaping Technology / Building Society: Studies in Sociotechnical Change,* edited by Wiebe E. Bijker and John Law. Cambridge, MA: MIT Press, 1992.

Law, John. "Technology and Heterogeneous Engineering: The Case of Portuguese Expansion." In *The Social Construction of Technological Systems: New Directions in the Sociology and History of Technology*, edited by Wiebe E. Bijker, Thomas P. Hughes, and Trevor Pinch. Cambridge, MA: MIT Press, 2012.

Lawrie, Paul R. D. *Forging a Laboring Race: The African American Worker in the Progressive Imagination*. New York: New York University Press, 2016.

Le Goff, Jacques. "Labor Time in the 'Crisis' of the Fourteenth Century: From Medieval Time to Modern Time." In *Time, Work, and Culture in the Middle Ages*, translated by Arthur Goldhammer. Chicago: Chicago University Press, 1980.

Le Hodey de Saultchevreuil, Étienne, ed. *Journal des États généraux*. Paris: Chez Devaux, 1789.

Le Peletier de Saint-Fargeau, Félix. *Réflexions sur le moment présent, offertes à la Convention Nationale, par F. Lepelletier*. Paris: Imprimerie de R. Vatar, 1793.

Lears, T. J. Jackson. *Rebirth of a Nation: The Making of Modern America, 1877–1920*. New York: HarperCollins, 2009.

Lécuyer, Christophe. *Making Silicon Valley: Innovation and the Growth of High Tech, 1930–1970*. Cambridge, MA: MIT Press, 2007.

Lefevre, Wolfgang. *Picturing Machines 1400–1700*. Cambridge, MA: MIT Press, 2004.

Lemann, Nicholas. *The Promised Land: The Great Black Migration and How It Changed America*. New York: Vintage, 1992.

Lemov, Rebecca. "Hypothetical Machines: The Science Fiction Dreams of Cold War Social Science." *Isis* 101, no. 2 (2010): 406–407.

Lenotre, G. *La Guillotine et les exécuteurs des arrêts criminels pendant la révolution, d'après des documents inédits tirés des Archives de L'état*. Paris: Perrin & Cie, 1920.

Lepore, Jill. *The Mansion of Happiness: A History of Life and Death*. New York: Vintage, 2013.

Lerman, Nina E. "Categories of Difference, Categories of Power: Bringing Gender and Race to the History of Technology." *Technology and Culture* 51, no. 4 (2010): 893–918.

Lerman, Nina E., Arwen Palmer Mohun, and Ruth Oldenziel. "Versatile Tools: Gender Analysis and the History of Technology." *Technology and Culture* 38, no. 1 (1997): 1–8.

Leslie, Stuart W. "Blue Collar Science: Bringing the Transistor to Life in the Lehigh Valley." *Historical Studies in the Physical and Biological Sciences* 32, no. 1 (2001): 89–90.

Li, Tania Murray. "Governmentality." *Anthropologica* 49, no. 2 (2007): 275–281.

Licht, Walter. *Getting Work: Philadelphia, 1840–1950*. Philadelphia: University of Pennsylvania Press, 2000.

Light, Jennifer S. "When Computers Were Women." *Technology and Culture* 40, no. 3 (1999): 455–483.

Lohan, Maria, and Wendy Faulkner. "Masculinities and Technologies: Some Introductory Remarks." *Men and Masculinities* 6 (2004): 319–329.

Long, Pamela O. *Engineering the Eternal City: Infrastructure, Topography, and the Culture of Knowledge in Late Sixteenth-Century Rome*. Chicago: University of Chicago Press, 2018.

Losurdo, Domenico. *Liberalism: A Counter-History*. New York: Verso Books, 2014.

Lott, Eric. *Love and Theft: Blackface Minstrelsy and the American Working Class*. New York: Oxford University Press, 1993.

Louis, Antoine. "Sur les charlatans." Antoine Louis Papers, Bibliothèque de Metz, Réserve Précieuse MS 1310.

Louis de Lacretelle, Pierre. *Discours sur le préjuge de peines infamantes*. Paris: Cuchet, 1784.

Lusser, Robert. "The Notorious Unreliability of Complex Equipment: A Condensed Presentation for Management Use." Huntsville, AL: Research and Development Division, Ordnance Missile Laboratories, Redstone Arsenal, 1956.

Lusser, Robert. "Unreliability of Electronics: Cause and Cure." Huntsville, AL: Research and Development Division, Ordnance Missile Laboratories, Redstone Arsenal, November 1957.

Macdonald, Stuart, and Ernest Braun, "The Transistor and Attitude Towards Change." *American Journal of Physics* 45, no. 11 (1977): 1061–1065.

MacKenzie, Donald. *Inventing Accuracy: A Historical Sociology of Nuclear Missile Guidance*. Cambridge, MA: MIT Press, 1993.

MacKenzie, Donald A., and Judy Wajcman. *The Social Shaping of Technology*. Cambridge, MA: MIT Press, 1999.

Mantel, Hilary. "He Roared." *London Review of Books* 31, no. 15 (2009).

Marcuse, Herbert. "Some Social Implications of Modern Technology." In *The Essential Frankfurt School Reader*, edited by Andrew Arato and Eike Gebhardt. New York: Continuum, 1982.

Marres, Noortje, and Javier Lezaun. "Materials and Devices of the Public: An Introduction." *Economy and Society* 40, no. 4 (2011): 489–509.

Marsden, Ben, and Crosbie Smith. *Engineering Empires: A Cultural History of Technology in Nineteenth-Century Britain*. Palgrave Macmillan UK, 2005.

Martschukat, Jürgen. "Ein Schneller Schnitt, Ein Sanfter Tod: Die Guillotine Als Symbol Der Aufklärung." In *Das Volk Im Visier Der Aufklärung: Studien Zur Popularisierung Der Aufklärung Im Späten 18. Jahrhundert*, edited by Anne Conrad, Arno Herzig, and Franklin Koptisch. Hamburg: Lit, 1998.

Marx, Leo. "Technology: The Emergence of a Hazardous Concept." *Social Research* 64, no. 3 (1997): 965–88.

Matt, Susan J. "Current Emotion Research in History: Or, Doing History from the Inside Out." *Emotion Review* 3, no. 1 (2011): 117–124.

Matus, Jill L. "Trauma, Memory, and Railway Disaster: The Dickensian Connection." *Victorian Studies* 43, no. 3 (2001): 413–436.

Mayr, Otto. *Authority, Liberty, and Automatic Machinery in Early Modern Europe*. Baltimore, MD: Johns Hopkins University Press, 1986.

Mbembe, Achille. *On the Postcolony*. Berkeley: University of California Press, 2019.

Mbembe, Achille, and Janet Roitman. "Figures of the Subject in Times of Crisis." *Public Culture* 7, no. 2 (1995): 323–352.

McIlwain, Charlton D. *Black Software: The Internet and Racial Justice, from AfroNet to Black Lives Matter*. New York: Oxford University Press, 2019.

McKittrick, Katherine. *Dear Science and Other Stories*. Durham, NC: Duke University Press, 2021.

Mead, Margaret. *Coming of Age in Samoa: A Psychological Study of Primitive Youth for Western Civilization*. New York: HarperCollins, 1928.

Medina, Eden. *Cybernetic Revolutionaries: Technology and Politics in Allende's Chile*. Cambridge, MA: MIT Press, 2011.

Meehan, Mary Beth, and Fred Turner. *Seeing Silicon Valley: Life Inside a Fraying America*. Chicago: University of Chicago Press, 2021.

Melamed, Jodi. "Racial Capitalism." *Critical Ethnic Studies* 1, no. 1 (2015): 76–85.

Mendon-Plasek, Aaron. "Mechanized Significance and Machine Learning: Why It Became Thinkable and Preferable to Teach Machines to Judge the World." In *The Cultural Life of Machine Learning*, edited by Jonathan Roberge, and Michael Castelle. Cham: Springer International Publishing, 2021.

Mercier, Alain. "Portefeuille Industriel: L'engrenage De La Prudence." *La revue* 18 (1997): 45–52.

Miles, Tiya. *All That She Carried: The Journey of Ashley's Sack, a Black Family Keepsake*. New York: Random House, 2021.

Mills, Charles W. *The Racial Contract.* Ithaca, NY: Cornell University Press, 2014.

Mindell, David A. *Between Human and Machine: Feedback, Control, and Computing before Cybernetics*. Baltimore, MD: Johns Hopkins University Press, 2004.

Mitchell, Timothy. *Rule of Experts: Egypt, Techno-Politics, Modernity*. Berkeley: University of California Press, 2002.

Mitman, Gregg. "Where Ecology, Nature, and Politics Meet: Reclaiming the Death of Nature." *Isis* 97, no. 3 (2006): 496–504.

Mohun, Arwen. *Risk: Negotiating Safety in American Society*. Baltimore: Johns Hopkins University Press, 2012.

Mohun, Arwen P., Thomas Le Roux, Tom Crook, and Mike Esbester. "Conclusion: Governing Risks in Britain and Beyond." In *Governing Risks in Modern Britain: Danger, Safety and Accidents, c. 1800–2000*, edited by Tom Crook and Mike Esbester. London: Palgrave Macmillan, 2016.

Monnier, Raymonde. *Le faubourg Saint-Antoine, 1789–1815*. Paris: Bibliothèque d'histoire révolutionnaire, Société des études robespierristes, 1981.

Moody, N. F. "The Present State of the Transistor and Its Associated Circuit Art." *Nuclear Instruments* 2, no. 2 (February 1958): 182–192.

Moody, N. F., and C. D. Florida. "Some New Transistor Bistable Elements for Heavy Duty Operation." *IRE Transactions on Circuit Theory* 4, no. 3 (1957): 241.

Moon, Francis C. *The Machines of Leonardo da Vinci and Franz Reuleaux: Kinematics of Machines from the Renaissance to the 20th Century*. Dordrecht, Netherlands: Springer, 2007.

Morgan, Henry. "Letter to the Editor." *Wireless World* 60 (1954): 178.

Morgan, Jennifer L. *Reckoning with Slavery: Gender, Kinship, and Capitalism in the Early Black Atlantic*. Durham, NC: Duke University Press, 2021.

Morgan, Philip D. *Slave Counterpoint: Black Culture in the Eighteenth-Century Chesapeake and Lowcountry*. Chapel Hill: University of North Carolina Press, 1998.

Morgan, Philip D. "Task and Gang Systems: The Organization of Labor on New World Plantations." In *Work and Labor in Early America*, edited by Stephen Innes. Chapel Hill: University of North Carolina Press, 1988.

Morus, Iwan Rhys. "Invisible Technicians, Instrument-Makers and Artisans." In *A Companion to the History of Science*, edited by Bernard Lightman. London: John Wiley, 2016.

Morus, Iwan Rhys. *How the Victorians Took Us to the Moon: The Story of the 19th-Century Innovators Who Forged Our Future*. New York: Pegasus Books, 2022.

Mrázek, Rudolf. *Engineers of Happy Land: Technology and Nationalism in a Colony*. Princeton, NJ: Princeton University Press, 2002.

Muhammad, Khalil Gibran. *The Condemnation of Blackness: Race, Crime, and the Making of Modern Urban America*. Cambridge, MA: Harvard University Press, 2010.

Mumford, Lewis. *Technics and Civilization*. New York: Harcourt Brace Jovanovich, 1934.

Mumford, Lewis. *The Golden Day*. New York: Horace Liverlight, 1926.

Nadel, Alan. *Invisible Criticism: Ralph Ellison and the American Canon*. Iowa City: University of Iowa Press, 1988.

Nakamura, Lisa. "Indigenous Circuits: Navajo Women and the Racialization of Early Electronic Manufacture." *American Quarterly* 66, no. 4 (2014): 919–941.

"Negroes Ought to Drive Plows, Not Autos Says Judge." *Chicago Defender*, May 11, 1912.

Nelson, Alondra. "Introduction: Future Texts." *Social Text* 20, no. 2 (2002): 1–15.

Nelson, Daniel. *A Mental Revolution: Scientific Management since Taylor*. Columbus: Ohio State University Press, 1992.

Nelson, Daniel. *Frederick W. Taylor and the Rise of Scientific Management*. Madison: University of Wisconsin Press, 1980.

Nelson, Daniel. *Managers and Workers*. Madison: University of Wisconsin Press, 1995.

Nelson, Daniel. "Scientific Management in Transition: Frederick W. Taylor at Johnstown, 1896." *Pennsylvania Magazine of History and Biography* 99, no. 4 (1975): 460–475.

Nelson, Daniel. "Taylorism and the Workers at Bethlehem Steel, 1898–1901." *Pennsylvania Magazine of History and Biography* 101, no. 4 (1977): 487–505.

Nelson, Daniel, and Stuart Campbell. "Taylorism Versus Welfare Work in American Industry: H. L. Gantt and the Bancrofts." *Business History Review* 46, no. 1 (1972): 1–16.

Nietzsche, Friedrich. *Menschliches, Allzumenschliches: Ein Buch für freie Geister*. Chemnitz: Verlag von Ernst Schmeitzner, 1878.

Noble, Safiya Umoja. *Algorithms of Oppression: How Search Engines Reinforce Racism*. New York: New York University Press, 2018.

"Noble Sees Reliability Failure Limiting War Role of ICBMs." *Electronics News*, November 21, 1960.

Norman, Benjamin M. *Norman's New Orleans and Environs*. New Orleans: B. M. Norman, 1845.

"Nouvelles Etrangères." *Gazette du Commerce*, August 26, 1775.

Nutch, Frank. "Gadgets, Gizmos, and Instruments: Science for the Tinkering." *Science, Technology, and Human Values* 21 (1996): 214–228.

Nye, David E. *America as Second Creation: Technology and Narratives of New Beginnings*. Cambridge, MA: MIT Press, 2003.

Nye, David E. *Technology Matters: Questions to Live With*. Cambridge, MA: MIT Press, 2006.

Oldenziel, Ruth. *Making Technology Masculine: Men, Women and Modern Machines in America, 1870–1945*. Amsterdam: Amsterdam University Press, 1999.

Olmsted, Frederick Law. *A Journey in the Seaboard Slave States, with Remarks on Their Economy*. New York: Mason Brothers, 1861.

O'Malley, Michael. "That Busyness That Is Not Business: Nervousness and Character at the Turn of the Last Century." *Social Research* 72, no. 2 (2005): 371–406.

O'Mara, Margaret. *The Code: Silicon Valley and the Remaking of America*. New York: Penguin, 2020.

O'Rourke, Stephanie. "Beholder, Beheaded:: Theatrics of the Guillotine and the Spectacle of Rupture." In *Visual Culture and the Revolutionary and Napoleonic Wars*. Edited by Satish Padiyar, Philip Shaw, and Philippa Simpson. New York: Routledge, 2016.

Orr, Julian E. *Talking About Machines: An Ethnography of a Modern Job*. Ithaca, NY: Cornell University Press, 1996.

Outram, Dorinda. *The Body and the French Revolution: Sex, Class and Political Culture*. New Haven, CT: Yale University Press, 1989.

Ozouf, Mona. *La Fête révolutionnaire 1789–1799*. Paris: Gallimard, 1976.

Paine, Thomas. *Rights of Man, Common Sense, and Other Political Writings*. Edited by Mark Philp. New York: Oxford University Press, 1998.

Parris, Henry. *Government and the Railways in Nineteenth-Century Britain*. London: Routledge & K. Paul, 1965.

Parsons, Talcott. "Evolutionary Universals in Society." *American Sociological Review* 29, no. 3 (June 1964): 339–357.

Pearson, Gerald L., and Walter H. Brattain. "History of Semiconductor Research." *Proceedings of the IRE* 43, no. 12 (1955): 1794–1806.

Perrow, Charles. *Normal Accidents: Living with High-Risk Technologies*. Princeton, NJ: Princeton University Press, 1999.

Perry, Claire. *Young America: Childhood in 19th-Century Art and Culture*. New Haven, CT: Yale University Press, 2006.

Perry, David. *Building the Public City: The Politics, Governance and Finance of Public Infrastructure*. Thousand Oaks, CA: Sage, 1995.

Peterson, Peter B. "Henry Gantt and the New Machine (1916–1919)." *Academy of Management Proceedings* 1986, no. 1 (1986): 128–132.

Peterson, Peter B. "The Followers of Henry L. Gantt (1861–1919)." *Academy of Management Proceedings* 1993, no. 1 (1993): 174–178.

Petiot, Linda. "Dirty Gertie: The DRTE Computer." *IEEE Annals of the History of Computing* 16, no. 2 (1994): 43–52.

Petrina, Stephen. "Sidney Pressey and the Automation of Education, 1924–1934." *Technology and Culture* 45, no. 2 (2004): 305–330.

Petryna, Adriana. *Life Exposed: Biological Citizens after Chernobyl*. Princeton, NJ: Princeton University Press, 2013.

Pickering, Andrew. *The Cybernetic Brain: Sketches of Another Future*. Chicago: University of Chicago Press, 2010.

Pickett, Steward T., and P. S. White. *The Ecology of Natural Disturbance and Patch Dynamics*. San Diego, CA: Academic Press, 1985.

Portal, Antoine. "Avis Important sur les moyens pratiqués avec succès pour secourir." Paris: L'Imprimerie Royale, 1776.

Portal, Antoine. *Observations sur la nature et sur le traitement de la phthisie pulmonaire*. Paris: DuPont, 1792.

Porter, Whitworth. *History of the Corps of Royal Engineers*. London: Longmans, Green, 1889.

Potofsky, Allan. "The Construction of Paris and the Crises of the Ancien Régime: The Police and the People of the Parisian Building Sites, 1750–1789." *French Historical Studies* 27, no. 1 (2004): 9–48.

Powell, John. *The Survival of the Fitter: Lives of Some African Engineers*. London: Intermediate Technology, 1995.

"President Taft Talks for Wilberforcians." *Baltimore Afro-American* (March 12, 1910).

Pritchard, Sara B. "An Envirotechnical Disaster: Nature, Technology, and Politics at Fukushima." *Environmental History* 17, no. 2 (2012): 219–243.

"Prix préposés par la Société établie à Londres pour l'encouragement des Arts des Manufactures & du Commerce." *Gazette du Commerce*, August 1763.

"Program Ninth Annual Session Southern Commercial Congress, New York October 15, 16 and 17." *Southern Good Roads*, October 1917.

Quigel, James Parvin, Jr. "The Business of Selling Efficiency: Harrington Emerson and the Emerson Efficiency Engineers, 1900–1930." PhD diss., Pennsylvania State University, 1992.

Rabinbach, Anson. *The Human Motor: Energy, Fatigue, and the Origins of Modernity*. Berkeley: University of California Press, 1992.

Ramelli, Agostino. *The Various and Ingenious Machines of Agostino Ramelli (1588)*. Translated by Martha Teach Gnudi. Baltimore, MD: Johns Hopkins University Press, 1976.

Rapport des commissaires chargés par le Roi, de l'examen du magnétisme animal. Paris: Imprimerie Royale, 1784.

Reddy, William M. "Sentimentalism and Its Erasure: The Role of Emotions in the Era of the French Revolution." *Journal of Modern History* 72, no. 1 (2000): 109–152.

Reddy, William M. *The Navigation of Feeling: A Framework for the History of Emotions*. New York: Cambridge University Press, 2001.

Reice, Seth. *The Silver Lining: The Benefits of Natural Disasters*. Princeton, NJ: Princeton University Press, 2001.

"Report of the Offices of the Railway Department, 1844–1845." *Parliamentary Papers* 39. London: HM Stationery Office, January 1846.

"Reports of the Inspecting Officers of the Railway Department." *Parliamentary Papers* 57 London: HM Stationery Office, 1861.

Restif de La Bretonne, Nicolas Edme. *Mes Inscripcions*. Edited by Paul Cottin. Paris: Librairie Plon, 1889.

Reynard, Pierre Claude. "Unreliable Mills: Maintenance Practices in Early Modern Papermaking." *Technology and Culture* 40, no. 2 (1999): 237–262.

Reynolds, Michael. *Engine-Driving Life*. London: Crosby Lockwood and Co., 1881.

Rhodes, Richard, ed. *Visions of Technology: A Century of Vital Debate About Machines Systems and the Human World*. New York: Simon and Schuster, 1999.

Rijke-Epstein, Tasha. "On Humble Technologies: Containers, Care, and Water Infrastructure in Northwest Madagascar, 1750s–1960s." *History and Technology* 37, no. 3 (2021): 293–328.

Riordan, Michael, and Lillian Hoddeson. *Crystal Fire: The Birth of the Information Age*. New York: W. W. Norton & Company, 1997.

Riquieti de Mirabeau, Honoré-Gabriel. *Mémoires biographiques, littéraires et politiques de Mirabeau écrits par lui-même, par son père, son oncle et son fils adoptif.* Edited by Lucas Montigny. Paris: Delaunay Libraire, 1835.

Riqueti de Mirabeau, Honoré Gabriel. *Travail sur l'éducation publique: Trouvé dans les papiers de Mirabeau l'ainé.* Edited by P. J. G. Cabanis. Paris: Imprimerie Nationale, 1791.

Riskin, Jessica. *Science in the Age of Sensibility: The Sentimental Empiricists of the French Enlightenment.* Chicago: University of Chicago Press, 2002.

Riskin, Jessica. *The Restless Clock: A History of the Centuries-Long Argument Over What Makes Living Things Tick.* Chicago: University of Chicago Press, 2016.

Robinson, Cedric J. *Black Marxism: The Making of the Black Radical Tradition.* Chapel Hill: University of North Carolina Press, 2000.

Rockman, Seth. *Plantation Goods: A Material History of American Slavery.* Chicago: University of Chicago Press, 2024.

Rodgers, Daniel T. *Age of Fracture.* Cambridge, MA: Harvard University Press, 2011.

Rodgers, Daniel T. *The Work Ethic in Industrial America 1850–1920.* 2nd ed. Chicago: University of Chicago Press, 2014.

Rohde, Joy. "Pax Technologica: Computers, International Affairs, and Human Reason in the Cold War." *Isis* 108, no. 4 (2017): 792–813.

Roosevelt, Theodore. "Fifth Annual Message." December 5, 1905.

Roscoe, Stanley N. *The Adolescence of Engineering Psychology.* Santa Monica, CA: Human Factors and Ergonomics Society, 1997.

Rose, Nikolas *Governing the Soul: The Shaping of the Private Self.* Stanmore, UK: Free Publishing, 1999.

Rose, Sarah F. *No Right to Be Idle: The Invention of Disability, 1840s–1930s.* Chapel Hill: University of North Carolina Press, 2017.

Rosenthal, Caitlin. *Accounting for Slavery: Masters and Management.* Cambridge, MA: Harvard University Press, 2018.

Rousseau, Jean-Jacques. *Oeuvres complètes*, vol. 3. Paris: Bénard et Compagnie, 1853.

Rousseau, Jean-Jacques. *Oeuvres complètes*, vol. 4. Edited by Bernard Gagnebin and Marcel Raymond. Paris: Gallimard, 1999.

Roze de Chantoiseau, Mathurin. *Tablettes Royales de Renommée.* Paris: Royer, Bailly, Veuve Duchesne, Desnos and Buisson, 1786.

Russell, Andrew, and Lee Vinsel. "After Innovation, Turn to Maintenance." *Technology and Culture* 59, no. 1 (2018): 1–25.

Saint-Edme, E. T. B. "Guillotine." In *Dictionnaire de la pénalité dans toutes les parties du monde connu.* Paris: Rousselon, 1828.

Sala-Molins, Louis. *Le Code Noir, ou le Calvaire de Canaan.* Paris, 1987.

Sandage, Scott A. *Born Losers: A History of Failure in America.* Cambridge, MA: Harvard University Press, 2006.

Saraiva, Tiago. *Fascist Pigs: Technoscientific Organisms and the History of Fascism.* Cambridge, MA: MIT Press, 2016.

Saraiva, Tiago. *The Orchard in Ruins.* Chicago: University of Chicago Press, forthcoming.

Satia, Priya. *Spies in Arabia: The Great War and the Cultural Foundations of Britain's Covert Empire in the Middle East.* New York: Oxford University Press, 2008.

Savot, Louis. *L'architecture françoise des bastimens particuliers.* Paris: François Clouvier and Pierre Aubouïn, 1673.

Sawday, Jonathan. *Engines of the Imagination: Renaissance Culture and the Rise of the Machine.* New York: Routledge, 2007.

Schaffer, Simon. "Easily Cracked: Scientific Instruments in States of Disrepair." *Isis* 102, no. 4 (2011): 706–717.

Schaffer, Simon. "Late Victorian Metrology and Its Instrumentation: A Manufactory of Ohms." In *The Science Studies Reader,* edited by Mario Biagioli. New York: Routledge, 1999.

Schatzberg, Eric. *Technology: Critical History of a Concept.* Chicago: University of Chicago Press, 2018.

Schivelbusch, Wolfgang. *Railway Journey: The Industrialization of Time and Space in the Nineteenth Century.* Berkeley: University of California Press, 1986. Reprinted in 2014.

Schlosser, Eric. *Command and Control: Nuclear Weapons, the Damascus Accident, and the Illusion of Safety.* New York: Penguin, 2013.

Schmidt, Tobias. "Adresse au Comité d'Agriculture et de Commerce de Assemblée Nationale de France," January 3, 1791. Archives Nationales de France, F17 1136.

Schnapp, Jeffrey T. "The Romance of Caffeine and Aluminum." *Critical Inquiry* 28, no. 1 (2001): 244–269.

Scott, James C. *Domination and the Arts of Resistance: Hidden Transcripts.* New Haven, CT: Yale University Press, 1990.

Scott, James C. "Everyday Forms of Resistance." In *Everyday Forms of Peasant Resistance,* edited by Forrest D. Colburn. New York: Routledge, 2015.

Scroggie, Marcus (a.k.a. Cathode Ray). "Transistor Graphic Symbols: A Critical Analysis of Existing Ideas and Conventions." *Wireless World* 63 (1957): 194–198.

Seigel, Jerrold. *The Idea of the Self: Thought and Experience in Western Europe Since the Seventeenth Century*. New York: Cambridge University Press, 2005.

Shapin, Steven. "The Invisible Technician." *American Scientist* 77 (1989): 554–563.

Shapin, Steven, and Simon Schaffer. *Leviathan and the Air-Pump: Hobbes, Boyle, and the Experimental Life*. Princeton, NJ: Princeton University Press, 1985.

Shapiro, Herbert. "The Impact of the Aptheker Thesis: A Retrospective View of *American Negro Slave Revolts*." *Science and Society* 48, no. 1 (1984): 52–73.

Simmons, Jack. *Express Train and Other Railway Studies*. Nairn: D. St. J. Thomas, 1994.

Sinclair, Bruce. "Integrating the Histories of Race and Technology." In *Technology and the African-American Experience: Needs and Opportunities for Study*, edited by Bruce Sinclair. Cambridge, MA: MIT Press, 2004.

Sinclair, Bruce, ed. *Technology and the African-American Experience: Needs and Opportunities for Study*. Cambridge, MA: MIT Press, 2004.

Slaton, Amy E. *Race, Rigor, and Selectivity in U.S. Engineering: The History of an Occupational Color Line*. Cambridge, MA: Harvard University Press, 2010.

Slayton, Rebecca. *Arguments That Count: Physics, Computing, and Missile Defense, 1949–2012*. Cambridge, MA: MIT Press, 2013.

Slayton, Rebecca, and Brian Clarke. "Trusting Infrastructure: The Emergence of Computer Security Incident Response, 1989–2005." *Technology and Culture* 61, no. 1 (2020): 173–206.

Smiles Samuel. *Self-help, with Illustrations of Character and Conduct*. London: John Murray, 1859.

Smith, Mark M. *Mastered by the Clock: Time, Slavery, and Freedom in the American South*. Chapel Hill: University of North Carolina Press, 1997.

Smith, Matthew Wilson. "Victorian Railway Accident and the Melodramatic Imagination." *Modern Drama* 55, no. 4 (2012): 497–522.

Smith, W. B. *The Color Line: A Brief in Behalf of the Unborn*. New York: McClure, Phillips and Co., 1905.

Smith, William. *Dictionary of Greek and Roman Antiquities*. New York: C. C. Little and J. Brown, 1853.

Sohn-Rethel, Alfred. *Das Ideal des Kaputten. Über neapolitanische Technik*. Bremen: Wassmann, 1926. Reprinted in 1990.

Soppelsa, Peter. "Paris's 1900 Universal Exposition and the Politics of Urban Disaster." *French Historical Studies* 36, no. 2 (2013): 271–298.

Soucy, Chester I. "A Broad Survey of the Military Electronic Equipment Reliability Problem." *Proceedings, 1956 Electronic Components Symposium*. New York: Engineering Publishers, 1956.

Soucy, Chester I. "Aerospace-System Applications of Reliability Technology Derived from the Electronics Field." In *Proceedings of the 4th Congress of the International Council of the Aeronautical Sciences*, edited by Robert R. Dexter. Washington, DC: Spartan Books, 1964.

Soucy, Chester I. "A Perspective View of Electronic Component Part Reliability." In *Proceedings of the Canadian Military Electronic Components Symposium*, April 11–12, 1961, Ottawa, Canada.

Soucy, Chester I. "Reliability Control of Electronic Equipment in Aircraft and Weapons Systems: General and Management Aspects." *Canadian Aeronautics Journal* 3, no. 7 (September 1957): 222–231.

Star, Susan Leigh. "Power, Technology and the Phenomenology of Conventions: On Being Allergic to Onions." In *A Sociology of Monsters*, edited by John Law. London: Routledge, 1991.

Star, Susan Leigh. "The Ethnography of Infrastructure." *American Behavioral Scientist* 43, no. 3 (1999): 377–391.

Staum, Martin S. *Cabanis: Enlightenment and Medical Philosophy in the French Revolution*. Princeton, NJ: Princeton University Press, 2014.

Stearns, Peter N., and Carol Z. Stearns. "Emotionology: Clarifying the History of Emotions and Emotional Standards." *The American Historical Review* 90, no. 4 (1985): 813–836.

Steele, George, and Paul Kircher. *The Crisis We Face: Automation and the Cold War*. New York: McGraw-Hill, 1960.

Stoler, Ann Laura. *Race and the Education of Desire: Foucault's History of Sexuality and the Colonial Order of Things*. Durham, NC: Duke University Press, 1995.

Stovall, Tyler. *White Freedom: The Racial History of an Idea*. Princeton, NJ: Princeton University Press, 2021.

Suchman, Lucy A. *Human–Machine Reconfigurations: Plans and Situated Actions*. New York: Cambridge University Press, 2007.

Taschereau, Jules Antoine. *Revue rétrospective, ou Bibliothèque historique*. Edited by Jules Antoine Taschereau. Paris: L'Imprimerie de H. Fournier Ainé, 1835.

Taylor, Astra. "The Automation Charade." *Logic* 5 (2018): 149–188.

Taylor, Charles. *Sources of the Self: The Making of the Modern Identity*. Cambridge, MA: Harvard University Press, 1989.

Taylor, Frederick Winslow. "A Piece-Rate System: Being a Step Toward Partial Solution of the Labor Problem." *Transactions of the American Society of Mechanical Engineers* 16 (1895): 856–903.

Taylor, Frederick Winslow. "Notes on Belting." *Transactions of the American Society of Mechanical Engineers* 15 (1893): 204–259.

Taylor, Frederick Winslow. "On the Art of Cutting Metals." *Transactions of the American Society of Mechanical Engineers Proceedings* 28, no. 2 (1906): 34–40.

Taylor, Frederick Winslow. "Shop Management." *Transactions of the American Society of Mechanical Engineers* 24 (1903): 1337–1480.

Taylor, Frederick Winslow. *Shop Management*. New York: Harper and Brothers, 1911. Reprinted in 1912.

Taylor, Frederick Winslow. "Success: A Lecture to Young Men Entering Business." *Bulletin of the Taylor Society* 9 (1926): 68–76.

Taylor, Frederick Winslow. *The Principles of Scientific Management*. New York: Harper and Brothers, 1919.

Taylor, James. "Business in Pictures: Representations of Railway Enterprise in the Satirical Press in Britain 1845–1870." *Past and Present* 189, no. 1 (2005): 111–145.

Taylor, Keeanga-Yamahtta. *From #BlackLivesMatter to Black Liberation*. Chicago: Haymarket, 2016.

The Influence of Railway Travelling on Public Health. London: Hardwicke, 1862.

Thompson, Emily. "Machines, Music, and the Quest for Fidelity: Marketing the Edison Phonograph in America, 1877–1925." *Musical Quarterly* 79, no. 1 (1995): 131–171.

Thompson, E. P. "Time, Work-Discipline, and Industrial Capitalism." *Past and Present* 38, no. 1 (1967): 56–97.

Tompkins, Kyla Wazana. "On the Limits and Promise of New Materialist Philosophy." *Emergent Critical Analytics for Alternative Humanities* 5, no. 1 (Spring 2016).

Trabaud, Jean-François. *Principes sur le mouvement et l'équilibre, pour servir d'introduction aux mécaniques et à la physique. Premier traité*. Paris: J. Desaint et C. Saillant, 1741.

Tresch, John. "Technological World-Pictures: Cosmic Things and Cosmograms." *Isis* 98, no. 1 (2007): 84–99.

Tresch, John. *The Romantic Machine: Utopian Science and Technology After Napoleon*. Chicago: University of Chicago Press, 2012.

Trotter, William, Jr., ed. *The Great Migration in Historical Perspective: New Dimensions of Race, Class, and Gender*. Bloomington: Indiana University Press, 1991.

Trouillot, Michel-Rolph. *Silencing the Past: Power and the Production of History*. Boston, MA: Beacon Press, 2015.

Tsing, Anna Lowenhaupt. *The Mushroom at the End of the World*. Princeton, NJ: Princeton University Press, 2015.

Turkle, Sherry. *Life on the Screen: Identity in the Age of the Internet*. New York: Simon and Schuster, 1997.

Turner, Jonathan. "The Defence Research Board of Canada, 1947 to 1977." PhD diss., University of Toronto, 2012.

Tyler, Henry Whatley. "The Official Reports and Returns of the Railway Department of the Board of Trade." *Quarterly Review* 3, no. 221 (1862): 1–40.

"Variétés." *Gazette nationale ou le Moniteur universel*, February 4, 1790, 141.

"Variétés." *Journal général de la cour et de la ville*, February 3, 1790, 4–5.

Vaughan, Diane. *The Challenger Launch Decision: Risky Technology, Culture, and Deviance at NASA*. Chicago: University of Chicago Press, 1996.

Veblen, Thorstein. *Engineers and the Price System*. New York: B. W. Huebsch, 1921.

Velho, Raquel, and Sebastián Ureta. "Frail Modernities: Latin American Infrastructures Between Repair and Ruination." *Tapuya* 2, no. 1 (2019): 428–459.

Verbeek, Peter-Paul. *What Things Do: Philosophical Reflections on Technology, Agency, and Design*. University Park: Pennsylvania State University Press, 2005.

Vernon, James. *Distant Strangers: How Britain Became Modern*. Berkeley: University of California Press, 2014.

Vernon, James. "What Was Liberalism, and Who Was Its Subject?; or, Will the Real Liberal Subject Please Stand Up?" *Victorian Studies* 53, no. 2 (2011): 303–310.

Verrips, Jojada, and Birgit Meyer. "Kwaku's Car: The Struggles and Stories of a Ghanaian Long-Distance Taxi-Driver." In *Car Cultures*, edited by Daniel Miller. Oxford: Berg, 2001.

Vickers, Brian. "Leisure and Idleness in the Renaissance: The Ambivalence of Otium." *Renaissance Studies* 4, no. 1 (1990): 1–37.

Vidal, Fernando. *The Sciences of the Soul: The Early Modern Origins of Psychology*. Translated by Saskia Brown. Chicago: University of Chicago Press, 2011.

Vieira, Ryan. "Rethinking the History of the Risk Society: Accident Reporting, the Social Order and the London Daily Press, 1800–30." In *Governing Risks in Modern*

Britain: Danger, Safety and Accidents, c. 1800–2000, edited by Tom Crook and Mike Esbester. London: Palgrave Macmillan, 2016.

Vieira, Ryan. "The Epistemology and Politics of the Accidental: Connecting the Accident's Intellectual and Cultural Historiography." *History Compass* 11 (2013): 227–234.

Vila, Anne C. "Beyond Sympathy: Vapors, Melancholia, and the Pathologies of Sensibility in Tissot and Rousseau." *Yale French Studies* 92 (1997): 88–101.

Vila, Anne C. *Enlightenment and Pathology: Sensibility in the Literature and Medicine of Eighteenth-Century France*. Baltimore, MD: Johns Hopkins University Press, 1998.

Vila, Anne C. "Introduction: Powers, Pleasures, and Perils of the Senses in the Enlightenment Era." In *A Cultural History of the Senses in the Age of Enlightenment*, edited by Anne C. Vila. New York: Bloomsbury Academic, 2014.

Vincent-Buffault, Anne. *The History of Tears: Sensibility and Sentimentality in France*. Translated by Teresa Bridgeman. New York: St. Martin's Press, 1991.

Voskuhl, Adelheid. *Androids in the Enlightenment: Mechanics, Artisans, and Cultures of the Self*. Chicago: University of Chicago Press, 2013.

Wailoo, Keith, Karen M. O'Neill, Jeffrey Dowd, and Roland Anglin, eds. *Katrina's Imprint: Race and Vulnerability in America*. New Brunswick, NJ: Rutgers University Press, 2010.

Wajcman, Judy. *Technofeminism*. Malden, MA: Polity, 2004.

Wajcman, Judy, and Donald A MacKenzie, eds. *The Social Shaping of Technology*. 2nd ed. Buckingham: Open University Press, 1999.

Wallerstein, Immanuel. *The Modern World-System IV: Centrist Liberalism Triumphant, 1789–1914*. Oakland: University of California Press, 2011.

Watters, Audrey. *Teaching Machines: The History of Personalized Learning*. Cambridge, MA: MIT Press, 2021.

Wenzel, Siegfried. *The Sin of Sloth: Acedia in Medieval Thought and Literature*. Chapel Hill: University of North Carolina Press, 1960.

West, Cornel. *American Evasion of Philosophy: A Genealogy of Pragmatism*. Madison: University of Wisconsin Press, 1989.

White, Shane. "'It Was a Proud Day': African Americans, Festivals, and Parades in the North, 1741-1834." *The Journal of American History* 81, no. 1 (1994): 13-50.

White, Shane, and Graham White. *Stylin': African-American Expressive Culture, from Its Beginnings to the Zoot Suit*. Ithaca, NY: Cornell University Press, 1999.

Whitmire, Ethelene. *Regina Anderson Andrews, Harlem Renaissance Librarian*. Urbana: University of Illinois Press, 2014.

Wilkerson, Isabel. *The Warmth of Other Suns: The Epic Story of America's Great Migration*. New York: Vintage, 2011.

Willey, Angela. "A World of Materialisms: Postcolonial Feminist Science Studies and the New Natural." *Science, Technology, & Human Values* 41, no. 6 (2016): 991–1014.

Williams, Elizabeth A. *The Physical and the Moral: Anthropology, Physiology, and Philosophical Medicine in France, 1750–1850*. New York: Cambridge University Press, 2002.

Winter, Marian Hannah. "Juba and American Minstrelsy." *Dance Index* 6, no. 2 (1947): 28–47.

Wise, M. Norton. "The Gender of Automata in Victorian Britain." In *Genesis Redux: Essays in the History and Philosophy of Artificial Life*, edited by Jessica Riskin. Chicago: University of Chicago Press, 2007.

Wishnitzer, Avner. *Reading Clocks, Alla Turca: Time and Society in the Late Ottoman Empire*. Chicago: University of Chicago Press, 2015.

Wittje, Roland. "The Electrical Imagination: Sound Analogies, Equivalent Circuits, and the Rise of Electroacoustics, 1863–1939." *Osiris* 28, no. 1 (2013): 40–63.

Wolin, Richard. *Heidegger in Ruins: Between Philosophy and Ideology*. New Haven, CT: Yale University Press, 2023.

Womack, Andrew. "Ridicule and Wonder: The Beginnings of Minstrelsy and New York." *Afro-Americans in New York Life and History* 36, no. 2 (2012): 86–115.

Wrege, Charles D., and Ronald G. Greenwood. "Frederick W. Taylor and Industrial Espionage: 1895–1897." *Business and Economic History* 15 (1986): 183–193.

Yi, Yuan. "Malfunctioning Machinery: The Global Making of Chinese Cotton Mills, 1877-1937." PhD diss., Columbia University, 2020.

Young Choi, Tina. "Writing the Victorian City: Discourses of Risk, Connection, and Inevitability." *Victorian Studies, 43*, no. 4 (2001): 561–589.

Zallen, Jeremy. *American Lucifers: The Dark History of Artificial Light, 1750–1865*. Chapel Hill: University of North Carolina Press, 2019.

Index

Page numbers followed by an *f* denote a figure.

Publisher contact:
The MIT Press
Massachusetts Institute of Technology
77 Massachusetts Avenue, Cambridge, MA 02139
mitpress.mit.edu

EU Authorised Representative:
Easy Access System Europe, Mustamäe tee 50,
10621 Tallinn, Estonia
gpsr.requests@easproject.com

Printed by Integrated Books International,
United States of America